With appreciation . . .

To my wife, Bev;
to my son, Rick;
to an editor's editor, Mike Hamilton; and
to Harriet Modler
    the most capable assistant any author could want.
Thanks more than words can say.

OTHER BOOKS BY JEFFREY G. ALLEN

How to Turn an Interview into a Job
(also available on audiocassette)

Finding the Right Job at Midlife

The Placement Strategy Handbook

The Employee Termination Handbook

Placement Management

Surviving Corporate Downsizing

The Complete Q&A Job Interview Book

The Perfect Job Reference

The National Placement Law Center Fee Collection Guide

The Perfect Follow-Up Method to Get the Job

Jeff Allen's Best: The Resume

Jeff Allen's Best: Get the Interview

Jeff Allen's Best: Win the Job

Complying with the ADA: A Small Business Guide to Hiring and Employing the Disabled

Successful Job Search Strategies for the Disabled: Understanding the ADA

The Resume Makeover

Jeffrey G. Allen, J.D., C.P.C.

John Wiley & Sons, Inc.

New York ▪ Chichester ▪ Brisbane ▪ Toronto ▪ Singapore

Library of Congress Cataloging-in-Publication Data:

Allen, Jeffrey, G., 1943–
    The resume makeover / Jeffrey G. Allen.
       p.   cm.
    Includes bibliographical references.
    ISBN 0-471-04618-3 (cloth) — ISBN 0-471-04624-8 (pbk.)
    1. Resumes (Employment)   I. Title.
  HF5383.A5633   1955
  808'.06665—dc20   94-34133

Printed in the United States of America

10 9 8 7 6 5 4

About the Author

Jeffrey G. Allen, J.D., C.P.C., is America's foremost employment attorney. For almost a decade, Mr. Allen was a human resources manager with small business employers or small divisions of major companies. This direct experience has been coupled with his employment law practice for the past 20 years. As a certified placement counselor, certified employment specialist, and professional negotiator, Mr. Allen is considered the nation's leading authority on resume preparation.

Mr. Allen is the author of more best-selling books in the employment field than anyone else. Among them are *How to Turn an Interview into a Job, Finding the Right Job at Midlife, The Employee Termination Handbook, The Placement Strategy Handbook, Placement Management, The Complete Q&A Job Interview Book, The Perfect Job Reference, The Perfect Follow-Up Method to Get the Job,* the popular three-book series *Jeff Allen's Best,* and, most recently, *Successful Job Search Strategies for the Disabled.* He writes a nationally syndicated column entitled "Placements and the Law," conducts seminars, and is regularly featured in television, radio, and newspaper interviews.

Mr. Allen has served as Director of the National Placement Law Center, Special Advisor to the American Employment Association, and General Counsel to the California Association of Personnel Consultants.

Contents

Introduction

With today's upheaval in the job market, you'd better know how to write a resume. It doesn't matter whether you've been downsized out of your job, or just need the world to see more accurately how accomplished you truly are. The more closely the written picture reflects the "true you," the better the results in your job search.

Why? Because today's managers can spot false information in a number of ways. Maybe the dates on the resume are too vague; perhaps there are too many unexplained gaps. Are you exaggerating your achievements or your skills? It could be something as simple as a misspelled word or punctuation errors. Some ghoulish human resources types take delight in the equivalent of pulling wings off flies—spotting the errors that make your carefully constructed resume collapse under its own inaccurate weight.

Since your resume is a self-marketing device, you should make yourself look as good as possible, within the boundaries of common sense and honesty.

The bottom line is you'd better *be* your resume—professional and real. That's why I wrote this book. After almost three decades in the employment field, I know what your resume must do to get you an interview. It's easy and what's more, it won't take much more than a day.

This is not just another book filled with resumes. It's a complete how-to manual. Over 50 before-and-after examples clearly explain do's and dont's in the formats preferred by employment pros.

Even more important is the one-on-one feedback you need. I'll personally make sure your resume is reviewed by an expert in the field. Here's how it works: Once your resume is completed, send it with the form at the back of the book and a self-addressed, stamped envelope to:

Resume Makeover Response
P.O. Box 34444
Los Angeles, CA 90034*

Within a week, you'll have it back. Not only will you know you've written it correctly, but you'll have professional assistance at no extra cost.

Many people know what kind of position they seek. But if you're uncertain what kind of job you want, or are considering changing careers, begin with Chapter 1, "What a Resume Is (and Isn't)," and Chapter 2, "Planning Your Resume." Use them for essential insight *before* starting the workbook.

Then, when you've written and proofed your resume, get it in the mail fast. I look forward to hearing from you.

*No resumes will be reviewed nor returned without the self-addressed, stamped envelope.

1

What a Resume Is (and Isn't)

When a potential employer first looks at you, 99 times out of a hundred he or she is looking at a piece of paper known as a *resume*. Initially, more job opportunities are lost because of a poor resume than any other single factor.

With so much riding on that word picture, it's amazing how many misconceptions there are about the resume. It is *not* a biography of your life. It is *not* a catalog of skills. It is *not* a job application.

It *should be* a well-structured, easy-to-read presentation of your own capabilities and accomplishments, short and to-the-point. Its purpose is likewise clear cut: to intrigue a prospective employer to the point where he or she invites you for an interview. That's it—no mystery. Strange, then, that so many mistakes have been made, so many career moves aborted all because people don't know how to put together this simple, but crucial document.

With so much riding on it, you'd better know how the game is played. Resume formats, what to include, what to leave out, where to put a key paragraph, how the document should look, even the paper it's printed on counts. The result should be one tightly written, perfectly clear portrait of you and your related work accomplishments, education, and skills—no more, and certainly no less.

CONSIDER THE PLIGHT OF THE HUMAN RESOURCES MANAGER

For every position, no matter how elaborate the screening process, there are two basic piles: "consider again" and "reject." Running the gamut from one-page notes to 12-page personal presentations, 99 percent of all so-called resumes wind up as rejects.

To see how the system works, consider one typically harassed professional: Gary Burdens. Burdens is a human resources manager at the Midwest headquarters for the #2 maker of pharmaceutical products. The corporation typically receives 40,000 to 50,000 resumes annually. Hires usually average 200 to 300 per year. Every hire requires an average of two to three interviews. That means of all the thousands of resumes pouring into headquarters, only from 400 to 900 will ever result in an interview . . . no more than one or two out of every 100 resumes in the stack!

When does Burdens read the 20,000 resumes reviewed by the average employment manager? Not during the day . . . he has meetings to attend, ads to place, and, of course, interviews with the few whose resumes cleared the first hurdle.

So he lugs an average of 100 resumes home nightly. By the time dinner is over and the kids are tucked in, he can count on maybe an hour or two of reading. How does he spot the fraction of an ounce of gold within the tons of slag in the resume pile?

He doesn't read—he scans. He checks to see if the individual has the qualifications and career interests that mesh with current job orders. Poorly prepared resumes are the first to be discarded.

One of the first things he looks for are *results*. Most people list job assignments. They know what they're paid to do. Few actually specify the results they produced. Seeing quantifiable results immediately alerts Burdens that he may have found a real live candidate.

POOR EXAMPLE

"Sold biotechnical products in a territory from Denver, CO, to San Francisco, CA."

GOOD EXAMPLE

"Increased sales of gastrointestinal medication 'Prevasic' in my territory by 20% over a two-year span."

Your resume with specific achievements helps you stand out, giving interviewers like Burdens a reason to place you in the active candidate pile.

YOUR RESUME MUST SHOW YOUR ABILITY TO COMMUNICATE CLEARLY

A good resume does more than just describe skills and achievements. By the way it's written, it helps Burdens form a picture of you. A rambling resume points out an individual who can't clarify his or her own thought processes. So it gets dumped.

On the other hand, if your resume is brief, clearly written, and interesting, it makes you stand out. Burdens has a positive picture of a person who can work well within his firm. He's interested, because you've come alive through your word picture.

The next question is, "If I'm not a good writer, should I do my own resume?" Having read thousands of resumes, I cast my own vote for the self-written version. Any good human resources person can spot the prefab variety from a resume shop, because it sounds canned and tinny. You don't need to be brilliant to create a winning resume—just be organized. I've outlined all the right steps in the upcoming pages. Follow along and you won't go wrong.

One of the first things I recommend is that you lead off with a strong summary of your background . . . a mini "Who Am I."

POOR EXAMPLE

"Demonstrated ability to understand a company's long-term strategic visions and goals, complex business rules and inherent constraints, in order to design, develop, document, implement, and maintain a flexible, comprehensive, cost-effective database structure and processes application software for new or major revisions to existing products or processes."

What *is* this woman saying? By the time a weary reader gets to the end of this sentence, all comprehension is lost.

WORSE YET

"This letter explains my experience which can be verified with the enclosed references. I believe that food service is truly an art—much like a dancer executes the perfect turn, an actor commands the stage or a painter applies his brush to canvas."

You could weep from the sensitivity. Seeing this hand-written, so-called resume on pink paper with stars and moons as illustration, a personnel type would be likely to wince and toss.

GOOD EXAMPLE

"Nine successful years with Hyatt Hotels Corporation, Eastern Region. Promoted from various management positions to an Executive Committee Member. Supervised 25 employees and 4 managers while overseeing the operations of Front Office, Reservations, Guest Services, and Telecommunications. Manager of the Year—Hyatt Arlington—1989."

Clean; concise; specific; this summary leaves no questions about the capabilities and expertise of the candidate.

YOUR RESUME IS A PERSONAL PRESENTATION OF HOW YOU VIEW YOURSELF

When you go to your closet in the morning you don't throw on an old sweatsuit if you're going to a business meeting, simply because it happens to be the first thing you find on the rack. Likewise, your resume can't be a hodgepodge of whatever happens to fall into your mind the moment you're hitting the computer, word processor, or typewriter. The key is editing.

In preparation for writing this book, I've reviewed hundreds of resumes. A foreign-born candidate for an executive housekeeping position with a major hotel chain sent in a 23-page "Resume Update USA" that included everything from background, nationality, marital status, education, and further education to a dozen "To whom it may concern" letters and half-a-dozen photocopied certificates. Nobody ever said the hunt for jobs was conducted on a level playing field. This earnest candidate did not learn the all-important lesson of correct presentation for the American marketplace.

Very simply, what you leave out can be as important as what you put into your resume. An effective resume carefully recreates your true, professional beauty. It should *not* be an unretouched photo—warts and all!

YOUR RESUME IS THE FIRST WORK PRODUCT YOU SHOW A PROSPECTIVE EMPLOYER

A good resume does more than simply describe where you've been and what you've done. By the way it's written, it actually shows how well you think and communicate. Consider it like the first assignment you show to your employer and treat it with appropriate significance.

Two pieces of communication came into a midsize legal firm in Northern California. The first was a standalone letter, the second a resume with cover letter.

POOR LETTER EXAMPLE

"Does your company need additional clerical support? My name is Stephanie Carples and I would like to make myself available to do any letter or documentation typing you may need. I have 15 years experience as a secretary with knowledge of both medical and legal terminology."

This letter was printed using a very light draft font, so it was difficult to read. There was no address, just a phone number in the middle of the page. The paper was drugstore standard, and the letter had no semblance of a professional look. Yet this person asked for a position conveying a legal firm's appearance to the outside world!

GOOD RESUME EXAMPLE

Qualifications

More than seven years of experience providing clerical and administrative support, including:

- Scheduling appointments, interpreting policies and procedures, answering heavy phones, and dealing extensively with the public.
- Proficient in the use of IBM PC, WordPerfect 5.1, Symphony 2.0, Professional File, and general office equipment such as: fax, LaserJet printer, copy machine, etc.

This woman's resume was easy to read and review. She had excellent qualifications, and her presentation did her justice. Yes, she got the assignment!

YOUR RESUME HELPS YOU FOCUS ON A POSITION, NOT JUST SETTLE FOR A JOB

At the same time you're creating a resume, hopefully to encourage companies to recognize your employment potential, you should also be selecting the kinds of positions you're willing to accept. You can create target resumes dedicated to your interest in a single industry or type of position. And you needn't limit yourself to just one. Considered as a marketing tool with a variety of formats, your resume has infinite possibilities, especially when you create more than one. Though no resume alone will win the job for you, without a good one you can't even proceed to the next level.

2

Planning Your Resume

Before you begin making over your resume or creating one for the first time, consider what your previous job hire experience has been.

FIRST MAKE SURE THE FAULT LIES WITHIN YOUR RESUME

- *Resume getting few or no responses*—Don't condemn your resume right away. It's possible you just don't *fit* into the position you're seeking. You may need further education, or retraining. On the other hand, maybe the fault lies squarely with your resume. In that case, you're in the right place!

- *Resume getting responses but no interviews*—Again, the problem may be something other than your resume. If you've gotten a nibble from an interviewer from a telephone chat but it leads to no first interview, the problem could be in your telephone technique. Or maybe the human resources representative sees through you, and she instinctively feels that you're not what you portrayed in your resume. That's all the more reason to make your resume reflect the *real* you—truthful and in command of your destiny.

- *Resume getting first interviews, but falling down in the clinches*—Once more, you may simply not fit in. If you've oversold or undersold yourself on paper, it comes through loud and clear during the initial interview. Review it for false advertising; check out your interview skills—and your deodorant.

- *"Always a bridesmaid, but never the bride"*—If your resume gets you in the door more than once, it's time to review your resume with a fine-tooth comb. When you're that close, there are usually very small differences that result in the offer being extended. If you can increase your chances by even one percent, it can make the crucial difference. A resume makeover can help by guiding the interviewer through your

6

accomplishments and emphasizing them in comparison to other candidates'.

IF YOUR CURRENT RESUME *ISN'T* WORKING, WHY NOT?

Sins of commission . . . sins of omission—there are so many things that can lead to the reject pile. Here are the ten most common:

- *Too long*—This is probably the most frequent error. You don't have to put everything down—it counts against you. With few exceptions, your resume is far more likely to attract favorable notice when you limit its content to *one* well-drafted page.

- *Disorganized*—With no recognizably consistent format, it's too hard to find the important information.

- *Overly wordy*—Are your sentences too long? Paragraphs too dense? Are you using three words when one precise one will do?

- *Lacks essential results or accomplishments*—In Chapter 1, I spoke about quantifiable achievements. Without them, you're missing strong selling points. You "saved your current employer *x* dollars with cost-cutting measures in office supply purchases." You "founded a specialty photographic laboratory and sold it just three years later for a substantial profit." You get the picture.

- *Too bare-bones*—Job-seekers who feel that "name, rank, and serial number" or employer, job title, and dates of employment constitute a full-blown resume are wrong. And they never get called.

- *Irrelevant information*—You can do more harm than good with data like age, height, weight, hobbies, even certain *telling* information about schools or organizations. Look at it this way: You're selling the professional aspect of your life . . . period. Employers don't want to know your golf handicap, and they have no right to other personal information. The rules have changed.

- *Unprofessional appearance*—The would-be secretary for the law office in Chapter 1 committed this sin. Poor typing, poor printing—even with the best qualifications, you can't get past a sloppy presentation.

- *Misspellings, poor grammar, and other gaffes*—I've met many personnel people who immediately consign a resume to the reject stack just because of misspelled words. It may not be fair, but it's real life. You should read the finished resume twice, and then have an individual whose spelling and grammar you respect proof it. That disinterested eye usually catches hidden errors.

- *The glitz effect*—Fancy binders are a waste of money and don't fit in manila files. Photographs and unusual paper colors are a deterrent rather than a benefit. If in doubt, leave it out. Stick to the proven formats in this book.

- *Misdirected to incorrect individuals*—Sometimes it's not the human relations department to whom you should be sending. It might be the department head.

FEEL GOOD ABOUT YOURSELF . . . AND WRITE A GREAT RESUME

If you're discouraged when you begin writing your resume, the results will show through. It may be difficult to psych yourself up when you're feeling lousy about the promotion you didn't get, or the pink-slip you just received. With some time for reflection, though, you can do it. The trick's in knowing your own value-added factors—what you, uniquely, can offer an organization. Here are some examples:

- Innovative techniques you've applied to solve key operating problems.
- Fresh, original ideas to improve the status quo.
- Proven ability to motivate others—the leadership factor.
- Knowing how to pare down expenses—real important in these lean times.
- Profit improvement methods—another surefire plus.

Employment mangers are looking for the value-added factor. If you can demonstrate that you have it, they'll want to see you because most mangers believe that such accomplishments predict future success. And they *need* you to be successful after hire to validate their own positions!

In the next chapter, as you review your background, you'll have a format for discovering and expanding on your own personal value-added factors.

TECHNIQUES TO MAKE YOU BEST ON PAPER . . . BEST IN PERSON

Looking at that blank piece of paper or computer screen can be very intimidating. One of the most important tricks to get past it is to pretend you're an interviewer and ask yourself the following series of questions:

1. *How do you stand out from the mediocre worker?* If you're truly committed to your career, you see yourself as above the competition. Exploit the positive differences you see.
2. *What specifically did you do in a given activity to prove your value?* A public relations professional wrote that she "planned and executed special events for airline account which earned an award from the Public Relations Society of America." An English teacher who went

abroad to teach Japanese students indicated that "In working with 32 public schools in Nagasaki, Japan, I wrote and delivered one motivational speech a week." It is these kinds of specifics that enable employment managers to see you in the flesh, even when all they're looking at is a piece of paper.

3. *Once you start outlining a series of bare facts, ask yourself "so what?"* "What difference will this make to the person reading my resume?" "Was what I did bigger or better than anyone else's?" "How did my role in persuading the Board of Directors to adopt a five-year management plan following a series of crises subsequently improve the commercial brokerage I worked for?" On the other hand, if you follow this train of thought for a given task and it leads to no particular advantage for you, leave it off the resume. Remember—you should plan to file no more than a single page.

4. *Envision yourself in a series of different scenarios.*

   a. Imagine that a good friend of yours works for an employer you'd love to be associated with. Imagine the two of them talking about you. What positive characteristics would your friend use to describe your benefits to the potential employer?

   b. It's your first day on the job, and you love it. What is it you love about this new position, and why do your new business associates appreciate you?

   c. Imagine yourself to be a visiting efficiency expert. You have to tell your employer what people can do to make themselves better at their jobs. Be very specific, as it relates to positions you've held. Once you've described what helped to make you very good at your job, it's great resume material!

Once your resume's complete, you can use these same methods to improve your in-person skills. Use them to enhance your ability to discuss your aptitudes and achievements when you're across the desk from the interviewer.

HOW TO MAKE WRITING YOUR RESUME AS SIMPLE AS FILLING IN THE BLANKS

Writing your resume is a series of decisions about what to tell of your related work experiences. Next chapter, you can let your thought processes flow freely as you jot down details in several categories. Through this broad information gathering you'll have the raw material you need to graft the vital elements into your chosen resume format.

3

Information Blocks—
Fill in the Blanks

Resume writing has three major parts: (1) Content—"What do I include; what should I leave out?"; (2) Format—"What kind of resume works best for me?"; and (3) Style—"How do I sell myself best?"

Before you start writing, you need all the basic facts at hand. By reviewing and filling out the forms in this section, you'll have a good idea of what you need to know. No need to procrastinate till you find every missing fact. Start now—see what you need and make a list.

BASIC IDENTIFICATION

Up at the very top goes your name, in capitals. If working in a word-processing program, you can also **BOLD** the letters. Although you will see headings listed on various parts of the upper portion of resumes, I feel that having it centered makes it stand out best.

Do not write "Resume" or "Job Qualifications for" or any other title for this document. The employment manager or department head knows exactly what it is.

Next comes your address line, followed by city, state, and zip code in upper- and lowercase letters. Finally, list your telephone number. This sounds so basic you may wonder why I even mention it. I ran across the resume of a very senior computer professional with over 30 years' experience recently. The first thing that jumped out at me was this man had carefully listed his enormous expertise but had inadvertently left out his telephone number!

Another important note: Do *not* list your office number, if you are currently employed, unless you are certain that you can receive phone calls at work where you can close the door without arousing suspicion. Sometimes, listing that number raises the red flag that your boss knows your plans and *wants* you to leave.

In any case, employment managers work long hours and they're quite used to calling candidates at home. One way you can give very specific information regarding your telephone availability is to include it in a cover letter (see Chapter 7). You may also state that your answering machine is always available to take messages.

If you have a spouse or other person at home during the daytime, do *not* entrust them to receive messages concerning your business life from prospective employers. It doesn't look good if the individual can't take a decent message; and the call is too important to leave to the chance of winding up with an incorrect name or telephone number.

SAMPLE HEADINGS

JOHN A. DOUGHERTY
876 Lavenia Place
Los Angeles, CA 90034
(310) 555–8724

HELEN CHEEVERS
4256 Elmwood Drive
White Plains, NY 10602
(914) 555–9004
Office: (212) 555–4444

EDUCATION

Working your information blocks chronologically is easiest for the majority of people. Feel free to skip around if you prefer.

The first major event of your life that employers care about is education. In terms of values attached to it, there are several points to know.

1. *Generally, the farther you have progressed in your career, the less important your degree becomes.* Exceptions are in very technical fields where continuing education is a prerequisite for advancement.
2. *Education usually comes after your work experience in the resume format.* That usually places it toward the end of the page. Again, there is an exception: If you're looking for your first full-time job after college, education precedes any work experience. Even up to five years after graduation, if your academic credentials are more impressive than your work history, list them prior to your employment record.

SAMPLE EDUCATION BLOCKS

B.S., Claremont College, Pomona, California, 1987
Major: Electrical Engineering, Minor: Computer Science

B.A. Business Administration, Yale University, 1993
1993 Crewing for Whitbread Round-the-World Yacht Race
1991–1992 Peace Corps, Bombay, India
Taught English; aided in hygiene training for several settlements within
an hour of the city.

In short, if your desired profession is not represented by any work
experience immediately after graduation, place your education closest to the top of the resume.

Scientists and engineers with outstanding credentials *should* put their
education before work experience, particularly with advanced degrees from leading universities.

3. *The most commonly awarded degrees may be abbreviated—Ph.D., M.A.,
    M.S., M.B.A., B.A., B.S.* Since new academic degrees are proliferating
    as coursework changes, spell out lesser-known ones: Master of Hu-
    man Services, not M.H.S.

4. *Indicate coursework only if it relates to your specific job objectives.* Then
    use these two tests to see if it should be included:

    a. Does it enable the employer to get a broader understanding of
        your related background when you are new to a field?

        OR

    b. Does it show your specific preferences in key work areas?

5. *Only indicate grade point average if you had at least a 3.3 (B+) average on
    the 4-point scale.* (Note: Check your final transcript or college regis-
    trar, if in doubt.)

6. *Academically, what other credentials are important to show off?*

Dean's List	Special scholarships given by the university
Phi Beta Kappa	Inclusion in publications like *Who's Who in American Colleges and Universities*
Summa Cum Laude	Special academic awards for majors in your field
Magna Cum Laude	Honor Roll

7. *Always lead off with the highest academic degree attained.* If you're a college
    graduate, high school information is unnecessary. (Still, you should
    fill in all relevant high school information since you may need it for
    job applications.)

8. *List appropriate internships during your school career, extracurricular activities that point to a leadership role, and part-time work, if it relates to your career path.* If such work helped finance your way through school, this is especially relevant if you are a recent graduate. Don't give a laundry list of extracurricular pastimes; it could mean a negative response in terms of thinking of you as a less-than-serious student. For this aspiring journalist, a relevant list might include:

SAMPLE EDUCATION-RELATED ACTIVITIES

Editor—Daily Bruin, UCLA
Roving Reporter—Daily Bruin
InterCollegiate Debate Team
Treasurer—Political Science Club

9. *Show continuing education courses and professional seminars.* Edging out another candidate with similar work credentials sometimes hinges on the employer's perceptions of continued growth in your field. A plant manager's listing might look like this:

SAMPLE PROFESSIONAL DEVELOPMENT

American Association of Industrial Management courses
Personal Computers School (Lotus 1-2-3)

It is not necessary to indicate whether you or your firm paid for the training.

Education Worksheets

High School (name, city, state) _____

_____

When Graduated _____

Major Studies _____

Honors, Awards, Class Standing _____

Clubs and Extracurricular Activities _____

_____

_____

College, Undergraduate (name, city, state) _____

_____

When Graduated _____

Degree and Major Field of Study _____

Internships/Outside Work Related to Degree _____

_____

Honors/Awards/Class Standing _____

_____

Clubs/Extracurricular Activities _____

_____

Special Accomplishments _____

_____

College, Postgraduate/Professional (name, city, state) _____

_____

When Graduated or Dates Attended _____

Degree/Certificate/License _____

Internships/Outside Work Related to Degree _____

_____

Honors/Awards/Class Standing _____

_____

Clubs/Extracurricular Activities _____

_____

Special Accomplishments _____

_____

Other Special Training/Courses _____

_____

Skills Learned _____

Course _____ Date Taken _____

Licenses or Certificates Held _____

VOLUNTEER ACTIVITIES

Whether or not you use materials from this category will depend on your career status. If you're a woman returning to work after an extended hiatus as wife/mother/caregiver, volunteer activities are essential. If you're seeking employment in the nonprofit sector, such listings can show your interest in a designated field. For example, if you targeted the "Tree People," an urban reforestation program headquartered in Los Angeles, membership and volunteer work in the Sierra Club would hold definite crossover interest.

Begin with local clubs, PTA, Junior League, etc. Then list national organizations such as Girl/Boy Scouts, YMCA, and American Red Cross. In terms of an organization like the Sierra Club or Smithsonian, you are a member by virtue of a magazine subscription (though this is a tenuous connection, at best).

Be sure that you point out the relationship between the activities in which you participated, and the skills you used. For example, if you were Chair of the Publicity Committee for the local chapter of the YMCA, you would have writing skills, media interface knowledge, and possibly have carried out public speaking assignments. If you were treasurer of your

Little League organization, you could list which computer programs you used, and your strengths related to bookkeeping and financial administration.

Volunteer Activities Worksheet

Name of Organization _____

City/State _____

Dates of Participation _____

Activities (Full description of each) _____

_____

_____

Offices Held _____

_____

Skills/Knowledge Acquired _____

Name of Organization _____

City/State _____

Dates of Participation _____

Activities (Full description of each) _____

_____

_____

Offices Held _____

_____

Skills/Knowledge Acquired _____

PROFESSIONAL ORGANIZATIONS

If you belong to any associations linked directly to your career, it is important that they be listed. It shows peer recognition and your commitment, investing time over and above the workday. Participation as an officer or awards that you won from competitions entered should be listed here. Membership cannot compensate for lack of work experience, but it shows you interacted with others in your field.

SAMPLE PROFESSIONAL ORGANIZATION LISTINGS

President: Independent Writers Organization
Los Angeles, CA

Recipient: James Edward Cross Award
Poetry Competition, 1992

OR

1980, State of Oregon, Certified Emergency Technician
1987, Sales Administration Certification, USDA

You may also belong to professional organizations not directly linked to your immediate job, but with stature in the marketplace; for example, as director of a child care center, you could be a vice president of your local Women in Business organization.

If the position you seek requires certification or licensing and you have not yet received it, write "Application Pending." Since requirements vary by state, check for your own area. Also make sure all information is completely accurate; locate copies and check exact date and name of accrediting agency.

SAMPLE PROFESSIONAL ACCREDITATION

1993 California State Bar Association Member No. 2357652

Professional Organization, Honors, Licensing Worksheets

Professional Organization Memberships

Organization _____

Dates of Participation _____

Offices Held _____

_____

Responsibilities _____

_____

Skills Acquired _____

Professional Organization Memberships

Organization _____

Dates of Participation _____

_____

Offices Held _____

_____

Responsibilities _____

_____

Skills Acquired _____

Honors

Honor _____

Awarding Organization _____

Date(s) _____

Honors

Honor _____

Awarding Organization _____

Date(s) _____

Licenses and Certification

License _____ When Awarded_____

License _____ When Awarded_____

Certification _____ When Awarded_____

PERSONAL INFORMATION

Personal information is now more a matter of what to leave out than what to include. In the past you would have included age, sex, marital status, number of children, health, hobbies, and sometimes even race and religion. Today, the personal data you put on the resume (way down near the end) is a calculated guess to see what it can buy you in terms of generating interviews.

Under all circumstances, you should leave off age, sex, race, and religion. It's against the law for the employer to ask, and it makes you look behind the times to use it. Besides, no use inadvertently playing to the employer's prejudices.

If you acquired your most recent advanced degree more than five years ago, leave off graduation dates. By the amount of time you've spent on career growth, savvy human resources people can figure your approximate age. Why make it easier for any problems to surface?

It's also useless to put "excellent health" because no one ever puts anything else, unless they're disabled. Then a whole other set of parameters comes into play.

What might you put on the resume *if* you're targeting a specific employer? Say you're single and you know the potential position involves a lot of travel. Revealing this information could be an advantage in this position. On the other hand, if you spot an opportunity at a traditional firm in a small town with a history of family values, indicating that you are "married with two children" may also be a benefit.

SPECIAL ABILITIES AND SKILLS

Use these to augment your experience and your education in several situations. Language skills are becoming increasingly important. If you speak Spanish in California, French in Quebec, or know a host of Asian languages, you may leap over the competition in fields like sales and hospitality.

Another example: With a strong math background and demonstrated success in studies such as computer simulation and modeling, data processing systems, algorithms, computer methods in statistics, and monetary analysis and policy, an inexperienced young graduate can earn respect and a host of interviews in the financial community.

You can present skills such as familiarity with desk-top publishing, photography, public speaking, and specialized writing background in concrete terms so that your value to the prospective employer is immediately understandable. Generally, such information follows the employment section.

On the other hand, functional skills such as planning and organizing abilities are not easily demonstrated. Rather than include these on your resume, save them for the body of your cover letter.

SAMPLE TECHNICAL SKILLS

Art and document conservation including washing, mounting, matting, framing, restoration

Extensive experience with the following desktop publishing programs: Aldus PageMaker, Adobe Illustrator

Special Abilities/Technical Knowledge Worksheets

Describe Target Position _____

Technical Knowledge Required _____

_____

List Areas of Knowledge Proficiency _____

_____

_____

How Did You Acquire This Knowledge? (Education, Hobbies, Other) _____

_____

Technical Skills Required _____

_____

List Areas of Skill Proficiency _____

_____

How Did You Acquire These Skills? (Education, Hobbies, Other) _____

_____

Describe Target Position _____

Technical Knowledge Required _____

_____

List Areas of Knowledge Proficiency _____

_____

How Did You Acquire This Knowledge? (Education, Hobbies, Other) _____

_____

Technical Skills Required _____

_____

List Areas of Skill Proficiency _____

How Did You Acquire These Skills? (Education, Hobbies, Other) _____

_____

GENERAL BUSINESS SKILLS

With the rapid expansion of the computer- and fax-driven office, simple bookkeeping and typing skills often no longer suffice. Even young beginners need a working knowledge of basic word processing and/or bookkeeping programs. Increasingly, professionals are being called on to do their own keyboarding. When I attended a WordPerfect class recently, a commercial real estate broker was there because her secretary was overwhelmed with work, so she was having to learn to do her own letters!

SAMPLE GENERAL BUSINESS SKILLS

Typewriter, adding machine, cash register, copier, fax machine, computer

SAMPLE ENHANCED SKILLS

WP 5.1—Highly Proficient; Microsoft Word—Strong Proficiency; Typing Speed 70+ wpm; Familiar with WordStar, Wang, Quattro Pro, Excel, and Lotus 1-2-3

Business Skills Worksheet

List Skill _____

How Used On-the-Job _____

List Skill _____

How Used On-the-Job _____

WORK EXPERIENCE AND ACHIEVEMENTS

While all the other elements in your resume play a strong supporting role, the experience section must be the star. Your background and accomplishments should grab the employer's interest, compelling him or her to interview you!

You get two big benefits from writing this exercise: better-quality leads and a systematic review of your work history so that you can verbally present your strengths during each interview. Even if you're fresh out of school and relying on your education, a strong presentation of employment experience shows future promise.

What's most current is what counts most. Spend the maximum amount of time on your current job; if unemployed, on your most recent position. Go back through your experience in reverse chronological order. Very little space should be spent on positions that go back more than 10 years, or on work that has little bearing on your present career direction.

As I noted in Chapter 1, it's not enough to list your work responsibilities if you expect your resume to do the selling job it's meant to do. It has to proclaim your specific value-added factors—results, accomplishments, the fruits of your labors. Where possible, add numbers or other quantifiable statements.

Can you match up your achievements to reflect such value-perceived factors as *motivation, courage, flexibility,* and *initiative?* How do your accomplishments show special knowledge, promotions, and commendations or other awards? Tying all these elements in with your specific duties positively shouts your worth to a prospective employer.

To keep the resume fast-paced, you should adhere to a specific writing style. Pronouns like *I, me, my* are out. What you want are brief, staccato phrases that begin with action-oriented verbs.

POOR EXAMPLE

My work included preparation and organization of production control data for review by management. I coordinated the flow of work between engineering and fabrication.

GOOD EXAMPLE

Planned and implemented annual comprehensive sitewide Total Quality Management (TQM) assessments and evaluations. TQM team achieved following goals for Laboratory: 1993—Gold Quality Award; 1991—Silver Quality Award.

Turn to Chapter 4 for a list of the best action-oriented verbs to spice up your resume writing. Make sure creativity doesn't outstrip accuracy. Include all necessary information about each of your most current positions: job title, dates, employer, type of product or service rendered (if not immediately known from employer's name), division, city, state, responsibilities, and special projects and accomplishments.

What should you do if you were part of a major team effort? As closely as possible, indicate your specific contribution in achieving the overall result.

Employment Experience/Achievements Worksheet

Photocopy this sheet and use a single page for each major employer.

Employer _____

Division (if any), City, State _____

_____

Goods Produced or Services Rendered _____

Position or Job Title _____

Primary Duties _____

_____

_____

Special Projects _____

_____

_____

_____

Accomplishments/Achievements _____

_____

_____

_____

Skills Learned _____

_____

REFERENCES

Do *not* include the names of references on your resume. It is completely unprofessional. Instead, at the bottom, use the following line: "Personal and professional references furnished upon request." If the job is a creative one where artwork or writing samples might be expected, you can say, "References and professional samples furnished upon request."

References are too precious to annoy, and you want to be able to contact them *first*, to let them know what to expect. The exception to the rule is for targeted resumes using highly motivated references within the target company. If you're relying on the endorsement of a current employee, make sure that person knows what you're doing before listing his or her name. Even so, it's better to refer to the reference in your cover letter.

More than ever, employers do check references. That's why it's essential that you select people who know you well, and who can speak favorably about your work habits, skills, and strengths.

Before you give out the name and number of a person you consider to be a positive reference on your behalf, ask his or her permission. Some people don't want the annoyance of phone calls from strangers, and they'll come through with a lukewarm presentation at best. Others are simply no good on the phone.

When you find a person who will accept this responsibility, be sure to give them a copy of your resume. Another helpful addition to their crib files is a series of questions you anticipate the interviewer may ask about you. (Be sure to include some helpful answers below the questions.) Stress that you don't mean for them to read the answers verbatim, but that such facts might prove helpful, when a sudden call comes in from one of your potential employers. Above all, make sure *you* are up-to-date on the latest phone number of the persons you are using as references. To have a prospective employer get a wrong number indicates you are sloppy and disorganized.

Reference Worksheet

Name _____

Professional or Personal Relationship to You _____

Title at Company/School _____

Address _____

Telephone _____

Name _____

Professional or Personal Relationship to You _____

Title at Company/School _____

Address _____

Telephone _____

SOLVING SPECIAL PROBLEMS

Military

How important this category is to you depends on how much time you devoted to military service. If you are completing 20 years of active duty and need to include it on your resume as a major part of your career, make sure you can parallel your duties with those in civilian life. For example:

Military Occupation	Civilian Equivalent
Supply Sergeant	Warehouse Foreman
Motor Pool Captain	Traffic Manager
Naval Communications Officer	Telecommunications District Manager
Military Police Sergeant	Security Manager

If, on the other hand, the military was a transitional part of your life, between high school and college, you need only account for the time lapse in your career. All you need to state is that you were a member of a specific branch of the military, that you completed your required tour of duty, and that you were honorably discharged. Like any other special training, you should mention any skills learned in the military that you can apply on the job.

SAMPLE MILITARY BACKGROUND

U.S. Air Force 1975–1979
Staff Sergeant
Obtained Welder's Certification

U.S. Citizenship

U.S. citizenship or proof of permanent residency is more important than ever to obtain work in the United States. The hue and cry against illegal aliens has meant that some companies may even go on witch-hunts in their overzealous care to employ only those legally able to work in this country.

Under normal circumstances, no mention of citizenship is needed on you resume. If you are a U.S. citizen and working toward a job with the federal government, you can simply answer affirmatively in your interview.

If you are not a legal resident of the Untied States, either by virtue of citizenship or the infamous *green card*, you'll find it difficult to find quality employment. For you, the best chance is to try to find a company that is willing to sponsor you for permanent residency status. Generally, this will hinge on your unique qualifications; in the past this has been the way large technical firms have acquired engineers and scientists from outside the Untied States.

If you lie to get a job and the truth comes out, the employer will not be kindly disposed towards you. The employer has invested time and money to train you. This time and money will be wasted should your ineligibility be discovered. A word to the wise: Do not expect the employer to rush to your defense—it is more likely that you will find yourself unceremoniously bounced from the company.

Security Clearance

Government clearances for classified work, like other special certificates and licenses, should appear prominently near the end of the resume.

Patents and Publications

If you're a technical type and responsible for an invention, discovery, patent, or other innovation that contributed in a major fashion to your field or your employer's success, mention it prominently near the top of your resume. It should be included as part of your summary of qualifications (more about this in Chapter 4).

Accolades received, papers presented at professional meetings, symposium proceedings published, articles in recognized journals, and anything else for which a scientist or educator receives acclaim can be meaningful in the job search.

Up to a full page of highlights—as an attachment to a one-page resume—is acceptable. If you have more you may wish to cut the list short with the notation: "Partial list only. Complete list of publications furnished upon request."

Gathering Lost Data

Because vagueness is the greatest single weakness found in most resumes, it's important that yours be as specific as possible. Get the dates, school courses, past jobs, and other facts you need before you begin writing. Look in your high school or college yearbooks, school catalogs, military certificate of service, school transcript, letters of commendation, newspaper clippings, and any other documents that provide facts.

If you are an older candidate, you might want to leave off telltale indicators of your age. That's fine; you need to be accurate, but you don't need to reveal everything.

SPECIAL CANDIDATES

If you fall into one of the following categories, you should pay special attention to these suggestions.

Newcomers/New Graduates

If you're a recent graduate, your education should precede your work experience. This is all the more important if you attended a prestigious university. Having said that, there are two main ways you can proceed with the body of your resume.

- *If you have relevant work experience*—Your resume can follow a reverse chronological format, like this budding marine biologist: Place your "EXPERIENCE" category next, and include any relevant internships, part-time jobs held during school, or summer full-time work. (See Chapter 4 for specific format.)

SAMPLE NEWCOMER CHRONOLOGICAL FORMAT

EDUCATION
University of California, Santa Barbara
B.S., Biology Major, 1993

Summer Studies: Catalina Island Sea Mammal Program, 1991

EXPERIENCE
Environmental Law Intern, Stevens, Parks & Shirley, Beverly Hills, CA
Assisted in preparation of pretrial files for clients in the environmental section. Summer, 1992

Arleta Aquatic Summer Camp, Tampa, FL
Head counselor. Instructed campers in scuba diving, snorkeling, lifesaving, and marine wildlife. Oversaw counselors whose campers

ranged from ages 13–16. Organized and coordinated all group activities. Summer, 1990

SKILLS AND ACTIVITIES
Marine Environmental Club
Proficient Microsoft Word, Lotus 1-2-3, LEXIS

- *If you don't have relevant work experience*—Check Chapter 4 for instructions on using the functional resume. In this format, you still list your education first, but you then follow it with categories including capabilities, accomplishments, and achievements. Focus on human factors that go beyond coursework, including leadership, computer skills, organizational ability. Talk about things you accomplished out of the classroom: the rally you coordinated on behalf of local political candidates, the rock concerts you promoted. At the end, briefly list your work experience. You should use a cover letter to further explain qualifications.

SAMPLE NEWCOMER FUNCTIONAL FORMAT

EDUCATION
Brown University, Providence, RI
B.A. Urban Studies, Grade Point Average 3.92
Summa Cum Laude, 1992

CAPABILITIES
- Write complete and detailed research reports
- Edit written materials for content and grammar
- Communicate effectively with librarians and others in research
- Experienced Internet user
- Generate charts and other visual aids to text on computer

ACHIEVEMENTS
- Edited college political magazine and wrote features on social issues
- Headed research team for Professor Samuel Lee's textbook on urban issues
- Won Senior Prize for essay on postriot revitalization in inner cities
- Created dormitory-based business delivering food for local restaurants

EMPLOYMENT
1991–1993—Dormitory Delivery Drivers; Organizer and Manager
Summer, 1992—Providence Democratic Committee; Campaign Worker
Summer, 1990–1991—Herald Plan Insurance; Telemarketer

Returning Warriors

The good news is that women returning to the job marketplace after an extended hiatus caring for their families are a familiar commodity. If you fit this category, you probably have extensive experience within community organizations, have held offices in various groups, and even done part-time work.

Did you earn a law degree and use it to do pro bono work while your children were young? Did you organize garage sales on a regular basis? Be sure to indicate your organizational and merchandising skills. For past employers, list position and length of time, without going into much detail. (If the work period was for less than a year, you can lump together various functions, i.e., sales clerk, telephone research, etc.) Check Chapter 4, use the functional format, and place your education at the bottom of the page.

One concern of most women returning to work is how to explain the years while not employed. There are two methods: One is to simply state your expertise in various categories, follow it by part-time employment, and then your education, without making any mention of the years away from work. Naturally, you'll have to explain what you did during this time in an interview. The better way is to explain your absence in a brief, special section on your resume, under "Personal."

SAMPLE "RETURNING WARRIOR" FUNCTIONAL FORMAT

SALES/MERCHANDISING
- Sold an average of 3 dozen handmade collars and scarves, bi-monthly in crafts bazaars throughout the Las Vegas area
- Used Intuit computerized bookkeeping program to track results
- Won "Best-of-Show" in five competitions for unique display methods

MANAGING
- Managed small boutique in owner's absence
- Planned and coordinated all details for local chapter Muscular Dystrophy Bike-A-Thon that ultimately netted $25,000
- Arranged Cub Scout outings for two years

CREATIVE SKILLS
- Designed and sewed all collars for bazaars
- Oversaw interior decor for three neighborhood houses

EDUCATION
University of Arizona, Tucson, AZ
B.A. Elementary Education

PERSONAL
Raised three children from 1971–1990

Older Candidates

Forty Plus is a national organization that assists individuals who are 40 years old and over to hone their job-seeking skills and find employment. This group and others like it have made the washed-up senior executive a thing of the past. Still, there's no denying that with downsizing a permanent part of the nineties' economy, it's more difficult for senior, better-paid candidates to quickly find a new job.

People over fifty with highly specialized technical skills are most in demand. Women with managerial backgrounds also seem to find work faster than unemployed males, seemingly because they are in short supply. Over the last few years, the mean age for newly hired managers has been around fifty-two. That means that of all managers hired, half are *above* the fifty-two mark.

As an older candidate there are several things you should not list on your resume: your age, the year you graduated, all your jobs (go back only as far as you need to make the best impression—typically no more than 10 to 15 years), any date like a patent or publication that classifies you as *older*.

There are two schools of thought on what kind of resume an older candidate should use. Personally, I favor the reverse-chronological, because that's what most employment managers are used to seeing. On the other hand, I recognize that if you are especially qualified for a certain kind of job, you may prefer a functional resume to focus on specific expertise. By stressing achievements, it puts the emphasis on what you've done, rather than where you've worked.

Be sure to use a summary following personal identification. It hones in on your experience in succinct form, and persuades the reader to read the rest.

SAMPLE CHRONOLOGICAL RESUME FOR SENIOR CANDIDATES

SUMMARY: Energetic Operations Manager. Master of Science degree, with over 10 years' experience supervising 50 employees, directing overall company administration and operation.

PROFESSIONAL EXPERIENCE

1987–Present—Manager of Operations
Johnson Brothers Manufacturing, Bridgeport, Connecticut
Manufacturers of wheels and axles for materials-handling equipment.

- Prepared a comprehensive customer service manual that reduced troubleshooting time by over 50%.
- Surveyed data-processing equipment and negotiated contract for purchase.
- Improved clerical work flow significantly, receiving the annual Most Valuable Employee Award in 1991.

- Interfaced with sales staff to reduce order-to-delivery time by an average of two weeks.

1981–1987—Senior Engineer
Allied Machine Parts, Inc., Hidden Valley, New York
Manufacturer of specialty valves for heavy-duty trucks.

- Developed prototype for a valve that became the leading product line.
- Monitored status of parts inventory, coordinating with purchasing, operations, and manufacturing departments to ensure adequate supplies.
- Supervised technical support personnel in absence of chief engineer.

AFFILIATIONS
Vice President: Connecticut chapter, Engineering Management Society
Member: American Association of Business Executives

EDUCATION
Registered Professional Engineer
M.S. in Business Administration, Princeton University, Princeton, New Jersey
B.S. Engineering, University of California at Los Angeles

REFERENCES
Personal and professional references available upon request, once mutual interest has been established.

SAMPLE FUNCTIONAL RESUME FOR SENIOR CANDIDATES

SUMMARY: Dynamic sales ability especially in opening new target markets. Strong closer. Imaginative coordinator for tie-in promotional campaigns. Managed over 100 East Coast accounts for large West Coast Gourmet Club.

SALES/BUSINESS TO BUSINESS
- Increased sales by $500,000 in 9-month period for Gourmet Club.
- Developed promotional campaigns for new product lines.
- Coordinated accounts at more than 20 large department stores.

SALES/RETAIL
- Increased profits by 25% at "Large-Size" boutique dealing with special and hard-to-handle clients.

MANAGEMENT
- Organized and implemented a program for 40 students abroad.
- Coordinated student/faculty liaison.

WORK EXPERIENCE

1985–Present   LE CELLIER, Whittier, CA
Outside and promotional sales

1983–1985   MORE TO LOVE BOUTIQUE, Studio City, CA
Retail salesperson

1981   LEARNING ADVENTURES ABROAD, Albany, NY
Student Activities Coordinator

EDUCATION
B.A. English, University of California, Berkeley

REFERENCES
Provided upon request

TOO MANY JOBS

What constitutes *too many* jobs on your resume depends on the field you're in. On the West Coast, large aerospace firms or job shops under contract to these firms would hire engineers, drafters, and other technical pros for specific projects. After two or three years, when the contract wound down, these people would be back on the market. Knowledgeable pros in human resources understood this phenomenon and so were not surprised to see an engineer with three or four different employers within 8 to 10 years.

In other industries where stability has been the norm, high job turnover is a negative in the minds of employment managers. Since reviewing resumes is basically a job of winnowing out undesirables, being classified as a frequent job changer is usually a negative.

How can you minimize what looks like a pattern of instability? One way is by using a functional resume, like the ones above, where you group skills by category. This can be further strengthened when you have standout achievements.

SAMPLE FUNCTIONAL RESUME FOR FREQUENT JOB CHANGER

Electrical Engineer
CAPABILITIES
- Conduct R&D for design, manufacture and testing of electrical components and systems
- Design manufacturing, construction and installation procedures
- Direct and coordinate field installation

ACHIEVEMENTS
- *Designed and drafted ship's electrical deck plans, including power, lighting, and security*
- *Designed and completed layout of plans, sections, and details of power distribution, systems, and security*

WORK HISTORY

1992–Present AGILON, INC. San Francisco, CA
  Junior Electrical Engineer

1990–1992 RAYTHEON CORP. San Mateo, CA
  Electrical Designer

1987–1990 HEWLETT PACKARD San Jose, CA
  Engineering Aide

Sometimes, there are a series of legitimate reasons why your resume looks a little more spotty than it should. In that case, you may want to indicate reasons for leaving. But be very careful: Some reasons are considered okay by human resource professionals; others just sound like wimpy excuses.

Among the preferred reasons for leaving: (1) Job elimination; (2) Acquisition by another firm that has placed one of its own employees in your position; (3) Nepotism, where the nephew of the boss moves into your spot; (4) Corporate downsizing; (5) Lack of advancement opportunity (use this one with care—it's often viewed as a disguise for general discontent).

Unacceptable reasons for leaving include: (1) Poor performance; (2) Incompatibility; (3) Absenteeism; (4) Dishonesty; (5) Quitting without providing the reasonable and customary two weeks' notice.

EMPLOYMENT GAPS

Where you have a work gap in your career pattern due to layoffs or underemployment, how do you handle it? If the period is less than a year, just leave it off, and blend in with your previous job. You do not have to list positions by months of employment; yearly will do fine.

If you were underemployed or out of work for more than a year there are two ways to handle it.

SAMPLE EMPLOYMENT GAP #1

1983–Present—Supervisor/Script Stock
IBM Corporation

1981–1982—Personal (Will discuss during interview)

1979–1981—Supervisor/Operations Department
Exmont Corp.

SAMPLE EMPLOYMENT GAP #2

1983–Present—Supervisor/Script Stock
IBM Corporation

1979–1981—Supervisor/Operations Department
Exmont Corp.

In the first case, you answer a question that may appear in the employment manager's mind. Since there is not enough information on which to base a negative opinion, he or she may opt to see you, provided your credentials are sufficiently compelling. On the other hand, if it's between you and an applicant without the mystery leave, you may not get the opportunity to discuss it.

In the second case, you simply omit the dates. The rationale is that the resume is not meant to provide every last detail, but to highlight the important segments of one's career. This may wash with you, but probably won't with the human resources officer. On the other hand, he or she may still be impressed with your background and wish to see you. There are no guarantees either way; you simply have to wing it to see which feels better to you.

4

Resume Styles— Which One Is Right for You?

You've put together your information. Based on what you've already read, you may have some vague inclination about general resume outline and format. Before you begin writing, it helps to get inside the employment manager's head. A few *do's* and *dont's* will move you in the right direction.

THE BIG *DO'S*

- **Get a name and a title for the envelope.** Envelopes with names get opened first. Call and get the correct spelling. Nothing annoys a person more than having his or her name misspelled. If possible, bypass the Human Relations Department by sending your resume to the person who wrote the requisition. Even if it's sent back to Human Resources, at least your name has registered with the individual who could ultimately be your boss.

- **Be picky and be specific.** As closely as possible, write your resume with a specific job objective in mind. You *won't* put the objective on the resume; it's too limiting. Still, your *unwritten* objective should include the position title, its responsibilities, potential, the industry, salary, and the region where you want to live. As you write, focus your thoughts on that prize job so that your resume is targeted in the right direction.

- **Think marketing!** Your target employer is the client; you're the commodity. Spend enough time to give your resume the polish it deserves. Clearly communicate how *you* meet the client's wants and needs.

- **Sell performance!** What can you do for this organization? Will you improve productivity? Increase profits? Cut costs or overhead? Create new products? Develop new services? Will you make the company better than before you were hired? How? Qualified applicants are everywhere. But employers want performers. That's what a standout resume can do for you—separate you from the pack and put you in the lead. Here are two examples for candidates seeking human relations positions:

BORING

Plan, establish, and conduct activities for the recruitment, screening, selection, and placement of all exempt employees for all branches. Supervise the maintenance and retention of employment files and records.

PRECISE, CONCISE, SPECIFIC

Increased productivity of professional employees by 25% by developing and administering extensive and ongoing management-by-objective programs. Established new HMO-based medical program that maintained employee satisfaction, while reducing costs by 15% quarterly.

- **Pay attention to timing.** Don't send your resume so that it hits on Monday (busiest catchup day) or Friday (notorious for terminations and exit interviews). Your resume should arrive between Tuesday and Thursday. If mailing locally, drop it in the mail on Monday morning. If it's going further, send it out on Friday for a Tuesday arrival.

- **If you've got a hot one on the line, don't waste time with the mails.** When you've already spoken with the potential employer and he or she is interested, spring for an overnight delivery service. Is this a better idea than faxing your resume? You bet. A fax looks muddy and can't possibly do justice to the effort you spent on creating and producing your resume. Above all, don't send your resume from work using the same overnight service your current employer uses. Somebody's sure to know and the results could backfire on you.

- **Follow up . . . follow up . . . follow up.** That doesn't mean to pester the firm to death. What it *does* mean is that you must keep track of every resume that you send; mark your calendar and call two or three days after you think it should have arrived. Don't fall prey to the "Don't call us; we'll call you" syndrome. You could wait forever. Confirm that they received it. Here's how a typical conversation might sound:

Them: Human Resources, Emma Waverly.

You: Good morning, Ms. Waverly. This is William Tucker. I sent you a resume last week for the _____ position. I'm calling to verify you received it.

Them: Just a moment Mr. Tucker . . . let me check. Yes, it's here. It's being reviewed by our professional development staff. (This is also a good time to learn the name of the person reviewing resumes.)

You: I have to decide on an offer, and I'd really like to talk with someone at EXXcom first. Could you give me the name of the person in charge of reviewing resumes, so I can find out if I will be interviewed in the near future?

Them: Yes, that would be Jim Farrell, extension 2345. He's department manager.

Of course, it won't always go so well. You could be met with "Don't call us. . . ." In that case thank the person as politely as possible. At least you have connected a pleasant personality to your name. But push no further.

- **Be brief and to the point.** I've said it before, but it bears repeating. Resume readership drops off drastically after the first page. Omit pronouns. Use bullet points to shorten and *punch up* your resume.

- **Use the language of your industry wisely.** If you're applying for work in the casino industry and don't know what the "cage" or "countroom" is, don't even think of using phrases like this in your resume. On the other hand, here are a couple of excellent paragraphs, targeted to specific readership:

MEDICAL CLAIMS MANAGER

Supervise Data Entry coding department. Work with Utilization Review area to ensure proper handling and timely processing of medical claims. Highly experienced in CPT-4 and ICD-9 coding procedures.

PROGRAMMER

Know COBOL, Assembler, PL/I, FORTRAN, JCL, Timesharing, Structured Programming, ISAM, VSAM. Responsibilities include application programming, statistical quality control, and microfiche conversion.

- **Numbers matter.** Use them where possible, like this radio sales manager:

  *Managed 29 people in 7 offices selling radio time for 15 stations.
  *Increased total billings from $3.2 MM to $10.5 MM.
  *Added 6 stations to listening audience in 2-year period.

Major *Do's* for Executive Resumes

Many executives fail to bring to their job search the competitive edge that they bring to their job. I've encountered many $100,000+-a-year executives who say, "I'm starting on this search right now. I don't want to waste time with a resume makeover. I'll just update my current one."

Current probably means at least five years old. The world changes in five years, and people with it. These same executives would never settle for resurrecting a five-year-old business plan to meet their companies' future objectives.

Not taking the time to do it right could easily mean months of rejection—and years of *rejection shock* that can knock any jobseeker cold.

- **Have an objective in mind.** If you're an upper-level executive, your objectives should include:
  - The name of the position.
  - The responsibilities and potential.
  - The type of company and industry.
  - The company size.
  - To whom you will report.
  - What you will report.
  - The salary you will receive.
  - Where you want to live.
  - What office environment is important to you.

  Write down your objective in one or two concise sentences. This is for you alone. Your resume won't state an objective, but practically everything on it should be aimed toward these goals.

- **A resume should answer the questions employers ask.** For executives, these questions increase in scope. Make sure your resume states:
  - How you can increase productivity.
  - How you can improve profits.
  - How you can enhance services.
  - Items that demonstrate your ability to perform.
  - That you know the employer's business.
  - That you understand the customers' needs
  - How you can make the hirer look good.

  Once you've targeted the jobs that meet your objectives, *your* resume is likely to meet the employer's objectives.

- **Adopt a market-driven approach that emphasizes results.** Sell the features (qualifications) and benefits (performance) of the product (you) and services (yours). Companies don't buy executives—they buy a stronger bottom line, a more effective marketing program, solutions to their problems. Know the problems they face and show how you can solve them.

- **Even if your career has been long and impressive, emphasize the most applicable, current, impressive achievements.**

- **Mention outside activities if they help.** An example might be management and leadership of a successful fundraising campaign.

- **Anticipate who will be reviewing your resume.** Executive-level jobseekers are often advised to direct their resumes to the highest appropriate official—CEO, CFO, COO, president, managing partner, or senior vice president. But recent studies suggest that executives at this level do not screen unsolicited resumes. Two-thirds of unsolicited resumes for executive-level positions received by senior officials are delegated for screening to employees at the director or manager level—usually in the human resources department. That information is helpful since you were a director or manager. You know how they think, what they've been told is important, and what they look for in a resume. If you're going to send an unsolicited resume, design it according to what a manager or director is conditioned to spot.

- **Use a chronological format that emphasizes dates.** When hiring high-priced talent, companies want to know what you've done recently, so emphasize current achievements.

- **Mention unusual experiences only if they drive up your value.** It's better to downplay the yearlong vacation for "self-discovery" you took. Employers think you might do it again. But two years in the Peace Corps might have taught you skills and insight useful to an employer. So would a yearlong executive management program at a business school. Consulting between corporate assignments can also be received favorably, if presented in the right way. Be prepared to answer the question, "If you were successfully self-employed, why do you want to return to a job?" You can explain that you prefer being a player instead of a coach.

- **Begin with a** *power summary.* I talked briefly about summaries for older employment candidates. Actually, they're essential for anyone with a decent background. For high-priced talent, they're critical. Use action language, quantifiable results, and a direct style to summarize properly.

POOR EXAMPLE

As the Public Relations Director for a large West Coast corporation, I have top-level corporate experience with heavy staff advisory, policy-building, and administrative responsibility. Excellent writing and speaking skills and strong editorial background are other attributes.

GOOD EXAMPLE

Director of Public Relations for Fortune 500 financial services corporation. Controlled budget of $5.6 million and staff of 40. Established media buying policy and revitalized corporate image.

Can you spot the problems in the first example? It's flat and it rambles. It talks about "experience" and "responsibility" without focusing on accomplishments. It lacks numbers. The second clearly communicates the candidate's level of experience with the designation "Fortune 500," and by stating the size of budget and staff control. The word "revitalized" suggests a turnaround that could interest the reader in the candidate. That's the purpose of a power summary—to motivate the reader to call you.

- **Follow your power summary with tangible achievements.**
- **Give a short synopsis of previous employers' statistics.** Whenever a current or former employer's size, sales volume, market share, or industry dominance can lend credibility to your credentials, make this information known.
- **Use the strongest words possible.** Not "responsible for, " but "directed," "managed," "controlled." (Top managers don't *coordinate*.) Then proofread and polish (more about this in Chapter 5).
- **Remember the one-page rule. Go to two pages only if you must.**
- **Finally, market test your resume.** Make ten copies of your final draft and distribute them to trusted friends and colleagues with exposure to top management. Ask for their honest evaluation of the resume's impact, clarity, and appeal. Then put the resume aside for one or two days before sending it out. Reread it again to make sure it profiles you properly.

THE BIG *DONT'S*

- **Don't update or emphasize in handwriting or typed strikeovers.** That means no underlining, no crossouts. Nothing should be added that hasn't been included at the time of your resume revision. If you don't have time for the complete revision, you're wasting your time sending the resume out at all.
- **Don't list references.** Use a brief statement at the bottom: "Personal and professional references furnished upon request." I explained the reason in Chapter 3. References should be notified in advance to expect a call, following a positive job interview. Having an employer call before they've even talked to you could nix the interview before it begins.
- **Don't state current salary or projected financial requirements.** It's a no-win gamble early on in the jobhunt. Too high and you're elimi-

nated. Too low and they wonder what's wrong with you. Besides, you want to have room to negotiate. When a prospective employer gets serious, it usually won't occur until the second interview. Even if you're asked for a salary range in response to a want ad, ignore it. Make your resume so strong that this lack of information won't count against you.

- **Don't include names of supervisors.** To do so shows lack of sophistication. If the employer wants this information she'll ask you for it. Your resume should contain information only about your abilities and skills; save the rest for the employment application.

- **Don't use an objective.** You see them all the time on resumes of the uninformed. Either they're so general they're useless, or they're so specific they limit the range of jobs open to you. Unless you have one very specific position in mind and are using a targeted resume and won't accept anything else, you lose more than you gain by including an objective. As I said earlier, you want to have that objective firmly planted in your mind so that your resume is written to reflect your wishes; you don't want to broadcast it. A power summary works much better.

- **Don't apologize.** If you don't have a college degree or all the qualifications advertised, never apologize or sound defensive in any employment communication. This includes resumes and cover letters. Accentuate the positive, even if you can't eliminate the negative. Do it well, and your audience will forget what you were supposed to have.

- **Don't enclose a photo.** Back in the bad-old-days before Equal Employment Opportunity laws, you could make a case for an attractive man or woman sending out a photo. It bespoke youth, enthusiasm, and so forth. Today, hiring on the basis of appearance (also translated as overt awareness of race or other personal characteristics) violates federal, state, and local equal employment opportunity laws, except under very limited circumstances like modeling or acting.

- **Don't reveal your age or race.** This more or less goes along with not sending a photo. However, in no way should this information appear in your "Personals" section. As I mentioned in Chapter 3, "Personals" are almost a thing of the past, except in isolated instances. You don't want to list when you graduated if it was more than five years ago. Listing certain colleges or organizations is also a clue not needed by the seasoned personnel pro. You might be proud, but they might be prejudiced.

- **Don't editorialize.** Report the facts in dynamic language; don't use adjectives. "Won three national writing awards for technical articles" makes your point much more concisely and objectively than "exceptional writing skills."

- **Don't mention firings or layoffs.** In Chapter 3, I spoke about people with an abnormally high number of jobs. Sometimes, in these special circumstances, there is a rationale for indicating a reason for leaving. (Review "Too Many Jobs" in Chapter 3.) Otherwise, leave it out—there's a strong stigma attached to "involuntary termination" and "layoff."

- **Don't be sarcastic, patronizing, or humorous.** Your tone should be quietly confident. An attorney friend of mine got a resume from a candidate saying words to the effect: "I know that given my experience this position is beneath me, but I need a job, so here's my resume." Save your political opinions for another forum. And unless you're applying for a gagwriting gig on one of the late-night talk shows, avoid humor. It won't be appreciated.

- **Don't exaggerate or mislead.** The average classified advertisement draws 200 resumes. While that may lead you to believe your facts won't be checked, be careful. Exaggerations in the experience portion of your resume can be easily checked. College degrees are among the facts that can be verified. If you left college just three units short of a degree, don't say you earned it. Rather, the education section of your resume should read:

  University of Nevada, Las Vegas; 4 years undergraduate study in hotel administration

  OR

  Drake University School of Law; 2 years

- **Don't send a cover letter to the human resources department.** A cover letter to an unidentified target can point you away from a position that's right for you. Those overworked personnel people will think of it as just one more piece of paper to shuffle. Worse yet, they may never look past it to your resume. You do want to send the cover letter to a departmental decision maker. (Read Chapter 6 for specifics.)

- **Don't use a resume writing service.** In Chapter 1, I mentioned that any personnel pro worth his or her salt can spot a resume from one of these sources because they all sound alike. When you make the effort to write your own resume, you prepare yourself for an effective job search. Discipline yourself to go through your past and select only that which truly matters. Then sit down and write copy that sells. It's the best practice for the rest of the search.

INCITING TO ACTION

Full sentences are unnecessary; they take up precious space. No pronouns like *I, me, my* are needed. It's understood that you're writing about yourself. For virtually all printed resumes, it's best to start a short sentence with an action verb—a word that presents a sharp image to the reader.

accelerated
accommodated
accumulated
achieved
acted
activated
adapted
adhered
adjusted
administered
adopted
advised
advocated
alleviated
analyzed
anticipated
applied
appraised
approved
arranged
ascertained
asserted
assigned
assimilated
assisted
assured
attained
audited
augmented
authorized

bought
brought
budgeted
built

calculated
chartered
checked
circumvented
classified
coached
collaborated
collected

communicated
compared
compiled
completed
composed
computed
conceived
concluded
conducted
conferred
confronted
consolidated
constructed
consulted
contributed
controlled
convened
converted
coordinated
correlated
counseled
created

dealt
decided
decreased
defined
delegated
demonstrated
designed
detailed
detected
determined
developed
devised
directed
disbursed
discharged
discovered
discussed
distributed
documented

earned

edited
elicited
eliminated
embellished
enacted
encouraged
enforced
engineered
enhanced
enlisted
enriched
entrusted
enumerated
equated
equipped
established
estimated
evaluated
evolved
examined
excelled
executed
exemplified
expanded
expedited
experienced
experimented
expounded
extracted

fabricated
facilitated
filed
financed
fixed
followed
forecasted
formulated
founded
furnished

gathered
gave
generated

got
governed
grappled
guided

handled
headed
helped

identified
illustrated
implemented
improved
improvised
inaugurated
increased
indexed
induced
informed
initiated
innovated
inspected
inspired
installed
instigated
instituted
instructed
integrated
interpreted
interviewed
introduced
invented
inventoried
investigated
isolated
issued

judged
justified

kept
keynoted
keyworded

lectured	persuaded	released	stimulated
led	pioneered	remunerated	strategized
listed	planned	rendered	streamlined
logged	practiced	renegotiated	strengthened
	prepared	reorganized	structured
made	presented	reported	studied
maintained	presided	represented	submitted
managed	prevented	researched	substantiated
manipulated	produced	responded	succeeded
mediated	programmed	restricted	summarized
memorized	progressed	reviewed	supervised
mentored	projected	revised	supplied
modeled	promoted		surpassed
motivated	provided	scheduled	surveyed
	published	screened	
negotiated		secured	taught
notified	queried	selected	theorized
	questioned	serviced	trained
observed		shaped	transformed
obtained	reasoned	signed	translated
operated	received	simplified	typed
ordered	reciprocated	simulated	
organized	recommended	solved	unified
originated	reconciled	specialized	updated
	recorded	specified	utilized
paved	reduced	spoke	
perceived	reevaluated	sponsored	validated
perfected	regulated	stabilized	verified
performed	reinforced	staffed	vindicated
perpetuated	rejected	standardized	wrote

THE CHRONOLOGICAL RESUME

What most people think is *the* resume format—where you list your employment record in reverse chronological order with your current or most recent job first—is not the only kind of resume around. It just happens to be the most popular.

Another name for it is "power resume," because it can power you to the top of the employer's list of must-see candidates. The most important information is up front—with the combination of a power summary and a dynamic presentation of your most recent experience. Using this format or its basic variation, the linear resume, usually generates the most interviews. But it's *not* for everyone.

When to Use the Chronological Resume

- When you have a consistent work history that relates directly to your next job target, without major employment gaps or numerous changes.
- When your job history shows consistent growth and development.
- When the name of the most recent employer is highly prestigious or important in your industry.
- When prior titles are impressive.
- When employers *expect* your resume to look this way, in such traditional venues as education, government, and "Fortune 500" firms.

When *Not* to Use the Chronological Resume

- When your work history is uneven, with gaps in employment.
- When you have had too many employers.
- When you have been doing the same thing for too long.
- When you have been out of the job market for an extended period.
- When you are seeking your first job after high school or college (unless you have related part-time experience).
- When you need to explain problems you dealt with in depth, sophisticated methods employed, complex solutions provided, and benefits produced.
- When your career goals change.
- When you have been an independent business consultant or a freelancer in the creative fields.
- When you want to deemphasize your age.

Basic Structure for the Chronological Resume

1. Start with the power summary. Use crisp, hard-hitting words to zero in on your most important points.

   Young, highly motivated Product Manager with excellent educational background and five years' experience with one of Los Angeles' top banks. Designed and implemented a statistical analysis program on salary expectations that accurately gauged future trends up to a year in advance. Developed and marketed new banking services for the senior management of banks in the $100–$300 million range.

   OR

Senior manufacturing executive with 20 years' responsible line and staff experience in medium-size metals manufacturing operation. Fast-track career growth achieved through bottom-line enhancement of corporate profits up to 15% annually. Take-charge personality who instills sense-of-purpose in my team.

2. Start with your present or most recent position, and work backwards. Devote the most space to your recent employment.

3. Give details for only the last five years; general information for ten years. Summarize early positions unless critical to target position.

4. Use year designations, not month or day.

5. Don't show every position with a given employer. List the most recent and two or three others, at most.

6. Within each position, stress major accomplishments. With your next job target in mind, emphasize those most closely related to that goal.

7. Education follows employment, unless you are a recent college graduate or are in the technical professions where a pedigree really counts.

8. Keep it to one page, unless there are *very* extenuating circumstances. For an experienced scientist, engineer, or senior executive, early drafts usually exceed one page. With so much vital information you may *have* to exceed a page. The problem is resumes are reviewed by someone who screens within a minute or so, and your rate of response diminishes by 20 percent for every page.

Joan Cummings
30 Arrow Road
Piscataway, NJ 08854
(201) 697-3452

Summary: Direct staff of 30 professionals in providing full range of Human Resource services to the Corporate Staff Headquarters complex. Designed and implemented fully computerized human resource data system estimated to save $500,000 yearly in time and increased productivity. Extremely strong management and motivational skills.

Experience:

Bensinger Manufacturing Company, Inc., Corporate Offices, New Brunswick, NJ
($850 million manufacturer of residential and commercial furnishings)

Director of Human Resources, 1987 - Present

Developed a computer model linking human resources planning with overall management and operations planning. Directed team of five in creating management incentive plan that attained a 20% increase in productivity during last fiscal year. Oversaw hiring of 90 engineers to support major capital expansion efforts. Functional responsibilities include: Human Resource Planning, Internal and External Staffing, Performance Evaluation, Salary Administration, and Training and Development.

Manager, Corporate Staffing, 1982 - 1987

Managed staff of 12 professionals and 13 support personnel, with annual budget of $1.5 million. Typical annual hires included 250 new hires and 300 internal placements. Personally worked with top management to conduct Director and Vice-Presidential searches. Reduced interview-to-offer ratio from 6 to 1 to 2 to 1, with savings of 45% in cost per hire. (Annual savings averaged $500,000.)

Spencer & Taub, Corporate Human Resources Consultants, New York, NY

Senior Consultant, 1979 - 1982

Designed and developed client employee evaluation programs. Successfully designed and installed three major systems based on clients' individual requirements.

Consultant, 1976 - 1979

Education:
M.B.A. Human Resource Mangement, Michigan State
B.A. Business Administration, UCLA
Magna Cum Laude

References: Personal and professional references provided upon request.

Ellis Voorzanger, Ph.D.
10523 Lost Creek Road
Northfield, MN 55064
(507) 819-3338

Summary: Scientist/Manager with 10 years' proven success in gene isolation, manipulation and analysis, and high-volume molecular marker analysis. Demonstrated leader in building new departments. Strong analytical skill and personal creativity have led to successful incorporation of newest technology into research and development projects: i.e., cloning from an agronomic plant species, establishing more rapid PCR assays for three major corn breeding traits.

Education:
Post Doctoral Research, Institute for Genetics, Cologne, Germany
(Cloned the transposable element Ds at the shrunken locus in corn)
Ph.D. Chemistry, Stanford University
B.S. Biochemistry, University of Illinois, Urbana

Professional Experience:
Biogenics Laboratories, Northfield, MN
Biotechnology and Data Analysis Consultant 1992-present
Devised a simple graphical method to map multigene traits. Wrote and submitted proposals to National Sanitary Foundation and Department of Energy to develop a computer program and create designer plants for bioremediation. Wrote and submitted proposal for Department of Labor to retrain defense workers in biotech/bioremediation.

Alviso Jennings Research Center, Stanton, MN
Department Head, Molecular Biology, 1987-1992
Corn Transformation team leader coordinating transgene analysis via PCR, Southerns, Westerns, and ELISA for timely selection of materials for advancement. Initiated the use of PCR as a replacement/augmentation for RFLP analysis. Optimized RFLPs, decreased costs, and increased lab throughput to 240,000 datapoints per year. Collaborated with domestic and international soybean, tomato, sweet and field corn breeders to establish and prioritize market projects based on market potential. Increased multigene breeding efficiency via improved genetic design.

Stauffer Chemical Co., Richmond, CA
Research Biologist, 1984-1987
Constructed vectors with first corn promoter (AD) returned to corn plants via transformation. Supervised one technician in cloning alternative strong corn promoters.

Professional Affiliation:
Representative, Midwest Plant Biotechnology Consortium, 1992 - Present

Skills:
Excellent knowlege of PC programs including Access, Improv, Quattro Pro, Lotus Write, WordPerfect, CorelDRAW!, Harvard Graphics, MathCad, Mapmaker, and Neural Ware.

References:
Personal and professional references will be provided upon agreement of mutual interest.

Drafting Your Chronological Resume

Following is a form to pull out the most relevant parts of your personal inventory from Chapter 3. Review the positions you want to include in your makeover resume—no more than five, preferably fewer.

Beginning with the most recent position on top of the drafting form, list everything related to your accomplishments on that job. Do the same with each position. (Photocopy extras for as many positions as you plan on using in your resume.)

Next, look over all the information you've listed; underline or highlight the activities or achievements that are most closely related to the next rung of your career ladder. At the bottom of the drafting form, rewrite the information you've highlighted into one concise, well-written paragraph that can be inserted in your resume. (See the samples that follow.)

Chronological Resume Drafting Form

Dates _____   Position_____

Employer _____

Achievement/Results for this position_____

_____

_____

_____

_____

Rewrite what you've underlined above in paragraph suitable for resume. _____

_____

_____

_____

_____

SAMPLE "BEFORE" PARAGRAPH

Ensured the vessels of the Victoria Clipper, an international fast ferry operation, departed safely and on time. In Canada, represented the company at the bargaining table where we negotiated a three-year union contract. Read consumer questionnaires from every passenger and responded to all concerns. In 1993, saved approximately $40,000 in telephone expenses through better line utilization and a commitment to our long distance carrier for one year.

SAMPLE "AFTER" PARAGRAPH

Coordinated proactive consumer-response system for Victoria Clipper ferry operation. Operated on maximum 5-day turnaround for all consumer mail and questionnaires. Supervised staff of three responsible for hourly on-time ferry departure. Represented management at bargaining table for three-year union contract; helped avert planned strike. Saved $40,000 in 1993 through enhanced telephone line utilization and negotiating yearly contract rates.

THE LINEAR RESUME—A CHRONOLOGICAL VARIATION

The linear resume is a 1990s version of the classic chronological resume. Instead of using a paragraph to explain job responsibilities and achievements, you itemize each significant fact on a separate line. Other than that it is basically a chronological resume, with the same advantages and disadvantages.

Linear Resume Advantages

- It feels comfortable to traditional employers, since it's just a variation of the chronological resume.
- It provides logical, easy-to-read flow.
- It is one of the easiest resumes to prepare.
- It emphasizes career growth, job progress, and continuity of employment.

Linear Resume Disadvantages

- Line-by-line listing takes up more room—a precious commodity when you're trying to limit the resume to one succinct page.

- It focuses on more current activity, when earlier experience may be better related to your new career objective. It shortchanges achievements that happened earlier in your career.
- It has the same disadvantages as the classic chronological resume in the case of too many or too few employers, when you have been out of the job market for an extended time, when you need to explain problems confronted and solved in depth, and when you want to deemphasize age.

Basic Structure for the Linear Resume

1. Following the heading, begin with a power summary, virtually identical to the one in the chronological section. Or, you may use a linear "Qualifications" format. Either way, it capsulizes your most important selling points.

   Summary: High-energy, results-oriented sales and marketing executive with over 12 years of proven achievements in all phases of sales and marketing. Established pattern of moving up employer brands into positions of market leadership.

   OR

   ### Summary of Qualifications
   - Dedicated to promoting atmosphere of professionalism in nursing
   - Committed, enthusiastic, people-oriented leader
   - Developed ongoing collaborative practice
   - Consistently balanced quality nursing care with substantive cost containment

2. The employment section follows, using the following categories:
   - *Employment description:* dates of employment, name of employer, location (city, state), company size, products, and services
   - *Your position:* title, length of time in the position (from . . . to), position defined quantitatively in terms of people and budgets managed, dollars impacted, your responsibilities
   - *Major achievements:* how the company benefited because you were there, dollars saved, earnings boosted, time efficiencies

3. Education—follows same format as chronological. Positioning depends on how long it's been since you graduated. More than five years—experience precedes; less than five years or in other than career-related activities, education precedes employment.

4. "Other" categories—might be military, community leadership, seminars (refer to Chapter 3).

5. References statement.

What the Linear Resume Should Look Like

- Dates go on the left side of the resume page, followed by company name and location employed. Both company name and location should be in capital letters and underlined; the company name itself should also be boldface if you have access to a computer or word processor with this feature.

- Title of position held goes on the next line. With the exception of initial capitals, the rest is in lowercase letters and the full title is underlined. Position itself should also be in boldface.

- If you have had more than one position with a single employer, separate it with a line of spacing. In this case, you would place the dates of employment for each position to the right of the position title, in parentheses. Doing this prevents any possible confusion with "employer" dates contained in the left margin.

- Follow with a brief description of employer and position—one line for each.

- Next comes three or four major achievements. If you are listing several positions, save the accomplishments for the top two. Use a bullet or asterisk and offset double spacing to make the achievement stand out. For earlier positions, a brief statement regarding position and responsibilities is sufficient. Note: With this format, it's imperative your accomplishments are truly worth noting; otherwise, select another resume style.

BRIAN JESSUP
2365 Holmes Avenue
Massapequa, NY 11758
(516) 987-0356

Senior quality assurance manager with 15 years' engineering, lab management, and inspection experience. Implemented annual cost savings ranging from 40K to 100K. Outstanding reputation as organizer and leader who consistently achieves objectives.

EXPERIENCE

1985
to
Present

ROUNDHILL FOOD SERVICE EQUIPMENT, Bedford-Stuyvesant, NY

Manager for this manufacturer of custom food service equipment

- Responsible for quality assurance, refrigeration/service, shipping/receiving, and National Sanitary Foundation (NSF) testing

- Design/review new product development

  - Reduced service calls by 60%
  - Substantially reduced "punch list" items
  - Estimated quality/purchasing cost savings of $75K per year

1977
to
1985

DONALDSON & CO. INC., College Point, NY

Quality Manager and Director of Quality Control for this manufacturer of commercial and professional refrigerators

- Established quality control department

- Supervised staff of 10 engineers and technicians

  - Implemented quality improvements with net cost savings of $100K yearly
  - Instituted computer program to detect fraud by outside service agencies

EDUCATION

Academy of Aeronautics, LaGuardia Airport, NY

Associate/Applied Science, Mechanical Technology, Queensborough Community College, NY

REFERENCES

Excellent references available upon request.

Linear Resume Drafting Form

Photocopy and use one such form per employer.

Dates _____ Employer _____

Location _____

Company Size, Products, Services_____

Position/Title_____

Position Size/Scope _____

Functional Responsibilities _____

_____

_____

_____

Major Achievements _____

_____

_____

_____

Rewrite in paragraph style suitable for insertion into your linear resume _____

_____

_____

_____

_____

THE FUNCTIONAL RESUME

Functional resumes have been the subject of dispute among resume experts for as long as I can remember. Some call them "Competency Cluster" resumes, or "Achievement" resumes. They group significant achievements or responsibilities together, irrespective of where they have occurred in your career. The problem with this is most employment managers prefer looking at a person's chronological progress. If you have been on the fast track or even steadily employed, you won't want to use a functional resume. However, there are instances where the functional resume can be truly versatile.

When to Use the Functional Resume

- When you want to emphasize capabilities not used recently.
- When you're just starting out after graduating high school or college.
- When you're reentering the job market after an extended absence.
- When your experience does not portray a clear-cut path of career growth.
- When you've had a variety of unconnected jobs.
- When you have been freelancing or consulting, or doing temporary work.
- When you're changing careers.

When *Not* to Use the Functional Resume

- When you want to show a consistent management growth pattern.
- When you're in highly traditional fields like politics, law, or the ministry.
- When you have a history of prestigious employers.
- Where you have performed only a limited number of functions in your work.

Basic Structure for the Functional Resume

1. Use the same straightforward heading as in any other resume, with name, address, city, state, zip code, and telephone number. All the same rules apply.
2. Use four or five separate paragraphs, each with a specific heading that describes one area of expertise or involvement. Or you may use a linear format.

3. Put the most important paragraphs closer to the top of the resume. For example, if you have been a freelance writer with editorial feature writing experience, public relations background, and advertising copywriting credentials, which of these you'd put closest to the top would depend on your current career direction.

4. Within each functional area, stress the most directly related achievements.

5. Recognize that you can include any relevant accomplishment without necessarily identifying the source of the experience.

6. Add a capsule version of work history, listing employer names and dates.

7. Include education at the bottom, unless it was within the last five years. Then it should precede your competency clusters.

8. Use the standard line about references being available upon request.

Selecting Functional Headings

Because there is greater flexibility in the functional resume format, it also means you can use a variety of paragraph categories. For example, you might feel your variety of work skills and functions merits attention. Include words and phrases like:

MANAGEMENT	LEADERSHIP	MARKETING
COMMUNICATIONS	TEAMWORK	SALES
TECHNICAL SKILLS	DESIGN	ADMINISTRATION
PROMOTION	SUPERVISION	QUALITY ASSURANCE
PLANNING	COUNSELING	FACILITATING
PURCHASING	INSPECTING	TESTING

Another option is to divide your functional resume by job-related categories. (This makes sense when you have held a variety of different positions. The closer you can come to linking the connection between jobs in your reader's mind, the better.) For example, a manager in a nonprofit organization might use the following list:

FUNDRAISING	PUBLIC SPEAKING	COMMUNITY AFFAIRS
PRESENTATIONS	PROMOTIONAL WRITING	NEEDS ASSESSMENT

Before you begin to write your functional resume, list all the functions that describe the depth of your experience. Then select the best four or five.

_____	_____	_____
_____	_____	_____
_____	_____	_____

George Madison
478 Aztec Lane
Phoenix, AZ 89457
(602) 443-0944

MANAGEMENT	Planned, budgeted, and managed development of products from inception through final production. Hired, developed, and supervised achievement-oriented professionals and technicians. Organized and coordinated teams of R&D marketing and market researchers for major projects.
TECHNICAL	Achieved technical and consumer objectives for leading professional hair and skin care products company. Solved critical problems in formulating shampoos, conditioners, moisturizers. Established first professional hair color "supplement" designed to be mixed with any dye to solve the major problems in hair coloring—fadage, improper coverage, and discoloration. Directed dry-skin treatment research.
COMMUNICATIONS	Maintained R&D liaison with marketing, market research, advertising, legal, and regulatory agencies. Presented key achievement and progress reports to senior technical and business management. Handled technical training programs for sales, marketing, and advertising staff and outside contractors. Prepared comprehensive safety and efficacy manuals to obtain management clearance for sale of products.

EMPLOYMENT

1982-Present   Guillaume For Hair & Skin, Phoenix, AZ
                     Group Leader

1980-1982   Elementals Pro Linea, Venice, CA
                     Senior Chemist

EDUCATION   B.S. Pharmacy, University of Georgia, Waycross, GA

REFERENCES   Available upon request.

Carlos Lacamara
6534 Edgeware Place
Fullerton, CA 93584
(714) 889-3661

Experience: Plant Foreman

- Supervised 15 machine operators, 10 assemblers, 3 set-up men

- Coordinated work force of 40 at maximum during 2 shifts

- Interviewed, hired, and trained all production labor

- Improved organization and safety of shop by analyzing needed repairs and maintenance

- Directed work on two press brakes, shears and other metal cutting, metal bending equipment

- Reduced downtime

- Reduced rejection record by improving quality of finished goods

- Assured on-time delivery to customers by regular interface with trucking lines

Work History:

Ralston Manufacturing, Division of SBX Corp., Cerritos, CA Manufacturers of materials handling equipment; $10 million annual sales, employs 40 plant employees

- Foreman of Fabrication and Assembly 1984-Present
- Machinist 1979-1984

Military:

U.S. Army   1974-1978
Worked with U.S. Army Corps of Engineers as welder

References available upon request.

Drafting Your Functional Resume

Use the following form to write one major segment of your employment, skills, or life experience. (As necessary, photocopy the form and use one each per significant heading.) Each heading will become a key paragraph in your functional resume.

Review your personal inventory in Chapter 3. On the top portion of the form, write down everything that relates to that single topic heading—*abilities, achievements, background*. Don't worry about style now; you're just looking for content.

Now, review what you've written and underline or highlight the most significant parts. In the lower part of the form, rewrite the materials into one cogent paragraph, using action verbs, concise language, and specific examples, where available.

Functional Resume Drafting Form

Key Paragraph Heading _____

_____

_____

_____

_____

_____

_____

Rewrite what you've underlined in paragraph or linear format suitable for resume use: _____

_____

_____

_____

_____

SAMPLE "BEFORE" PARAGRAPH

As junior drafter, I sketched precision surgical equipment including microscopic components. I got great training in very detailed work. I also got substantial background in the areas of electrical and optical materials. I began with "roughs" and proceeded to finished designs. Sometimes, I would accompany the engineer on special projects to do rough drawings and take specific measurements. Worked with computer.

SAMPLE "AFTER" PARAGRAPH

Prepared clear, complete, accurate sketches of precision surgical equipment. Accompanied engineers on site for special projects: translated these rough drawings and measurements into finished plans. Used CAD/CAM computer programs. Responsible for finished drawings used in production.

THE TARGETED RESUME

When you have your eyes on one very specific job prize, the targeted resume is for you. It focuses your search and because of its specificity, it is often called the "Achievement resume." In style, it speaks the language of a single industry or job range.

That's not to say you can't have several targeted resumes. In fact, you should. For example, if you were a computer programmer with experience at school district headquarters in a major city, you could write the same basic experience in two ways: one to continue on your career path at another educational facility, the other to move into the computer industry.

SAMPLE TARGETED PARAGRAPH FOR CAREER PROGRESS IN EDUCATION

Designed and implemented key aspects of Chicago's computerized college and career information system. Developed innovative work materials for students, counselors, and administrators. Supervised implementation at district's 26 high schools. Coordinated evaluation assessments; results showed a range of practical usage from 12% to 20% depending on specific project.

SAMPLE TARGETED PARAGRAPH FOR MOVE TO COMPUTER INDUSTRY

Designed and implemented key aspects of pilot and continuing phases of computerized information system. Scope included assessment of needs, costs, and benefits of various software and hardware options. Prepared varying levels of user materials. Integrated system into multiple locations. Arranged technical assistance and training on site.

When to Use the Targeted Resume

- When you are very clear about your next move.
- When you plan to apply in several fields and will need a special resume for each.
- When you want to emphasize capabilities that may not be directly related to any professional experience you possess.

When *Not* to Use the Targeted Resume

- When you want to use the same resume for a number of applications.
- When you are not clear about your skills and achievements.
- When you're just starting your career, with little experience.

Basic Structure for the Targeted Resume

1. Earlier in this chapter, I said don't use an objective. For every rule, there's usually an exception, and this is it. Right below your basic identification, instead of a power summary, put "Objective" or "Job Target." It should state in the clearest possible language a description of the position or field you are pursuing.

   Objective: Lead programmer with supervisory responsibilities

   OR

   Job Target: Assistant manager's position in large drugstore chain

2. Each statement of your ability and achievements should be directly linked to the target position.

3. Use a linear format to highlight your background. Tone should be very positive.

4. The capabilities section should answer the question, "What can you do?" The achievements section should answer "What have you done that worked?"

5. Work experience is provided as a series of listings, much like the functional format.

6. Education follows at the end.

7. Keep it to one page, with plenty of white space.

Hazel Fredericks
236 Post Street
San Francisco, CA 94234
(415) 878-9933

Job Target: Outside sales representative to the hotel industry.

CAPABILITIES:

- Self-starter; not afraid of long hours

- Tenacious at follow-up

- Proven strength in establishing new accounts

- Extensive knowledge of Bay Area hotels and key managers

- Consistent ability to meet sales quotas

- Computerized records keeping; able to quickly update on the road

ACHIEVEMENTS:

- Successsfully reacquired 800-room Shoreline Hotel as largest account

- Sold full line of hotel furnishings to the Hilton, Sheraton, and Radisson chains

- Worked with designers on specialty installations at the Ritz Carlton

- Targeted "boutique hotels" with antique furnishings that emphasize San Francisco's "heritage look"

- Consistently achieved #1 sales in my territory

EMPLOYMENT HISTORY

1987 - Present   Account Representative Herndon Furnishings, San Francisco, CA

1985- 1987   Trainee & Coordinator Matthews-Starr Glassware, Daly City, CA

EDUCATION

B.A. English, Iowa State University, Ames, IA

References available upon request.

Drafting Your Targeted Resume

To get the key elements of a targeted resume, you'll answer a variety of questions based on two key elements: your capabilities and your achievements. To begin, first think of three specific target jobs that would fit your background and interests. List them here:

Job Target #1 _____

Job Target #2 _____

Job Target #3 _____

Use the following format to identify your skills. Use the subsequent one for your achievements. Photocopy and use one of each per job target.

Capabilities Worksheet (What *Can* I Do?)

1. List five things you can do regarding this job target.

_____

_____

_____

_____

_____

2. What can you do to add value to this job target?

_____

_____

_____

_____

3. How can you be of financial value to this job target?

_____

_____

_____

4. Is there anything you could do that would produce consumer value for this job target?

_____

_____

_____

5. What could you do in a related field that might be of value to this job target?

_____

_____

_____

Achievements Worksheet (What *Have* I Done?)

1. List five results directly impacting this job target.

_____

_____

_____

_____

_____

2. What have you done to add consumer value to this job target?

_____

_____

_____

3. How can you add financial value to this job target?

_____

_____

_____

4. What have you achieved in allied fields that would be worthwhile in your target job?

_____

_____

_____

PUTTING YOUR RESUME TOGETHER

By now, you should know whether the chronological, linear, functional, or targeted resume meets your needs. You should already have written the key paragraphs or linear phrases that fit under the employment subheading. What else remains? Putting it all together. Before you begin, review these points:

1. **Use the clearest, most descriptive headlines for your information blocks.** Which works best for you? _Employment . . . Experience . . . Professional Experience . . . Work History . . . Executive and Professional Staff Positions_. Work History, Experience, or Employment sounds best for nonexecutive positions. Save Professional Experience or Executive and Professional Staff Positions when your background merits it.

Education is self-explanatory. But you may need subheads in this category. If appropriate, you may want to list _Representative Courses_. A new chemical sciences graduate might list Principles of Chemistry for Honors Students, Organic Chemistry, and Analytical Chemistry as significant courses. Or a nurse who has a bachelor's degree might want to use *Key Academic and Clinical Courses.*

List all other headings you'll need: *Honors, Publications, Military Service, Memberships, Organizations, Special Skills, Special Projects, Certificates, Licenses.*

2.   **Make all your headlines look consistent.** You may choose to underline them, capitalize them, place them flush left or right, or center them on the page. Just make sure each is positioned the same and uses the same methods of emphasis.

3.   **Make sure the key word of each statement is at or near the beginning of the line, and emphasize results as much as possible.** Clarity and accuracy count. Review the list of power verbs given earlier in this chapter. Use the strongest one to begin each statement. A social worker with a functional linear resume format might list:

Counseling
- Codeveloped a unique Bereavement Counseling Program for persons who have recently lost a loved one.
- Successfully engaged resistant clients in group counseling.

Using a paragraph format, this purchasing manager might write:

Directed activities of 15 employees. Purchased $75 million annually in chemicals, production equipment, supplies, and furnishings. Activated cost-cutting computer program responsible for savings of up to $450,000 annually.

4.   **Organize similar information graphically.** Here are several examples:

a.   Bullets help the reader understand individual subunits of an overall paragraph. As an alternative to asterisks, they are particularly useful for linear formats.

- Taught English as a Second Language for Spanish Speakers.
- Interpreted at international bilingual conferences.

b.   **Skill areas** can be indicated in boldface.

c.   <u>Job Titles</u> are all underlined.

d.   WORKPLACES are all in capital letters.

5.   **Make proper spacing work to your best advantage.** One powerful method uses columns, like the Work History in this functional-linear resume:

1985–present	Student, paralegal	UCLA, Certificate Program
1986–summer	Paralegal Intern	United Farm Workers, Fresno
1983–1985	Youth Counselor	Stepping Stones, Los Angeles

Here are some other spacing techniques:

- Managed Crisis Intervention team for major public relations clients:

    -Mobil Oil   -Hormel   -Colgate-Palmolive

OR

- Wrote foreign language articles and translated for clients in:
  . . . Spanish   . . . French   . . . Portuguese   . . . Japanese

6. **Where you have complex phrases, use action statements and break them up with subheads:**

TRAINING

- Trained hundreds of individuals at all levels in computer literacy:
  - corporate executives   ▪ accountants
  - sales clerks   ▪ office personnel
- Managed design and implementation of customized hands-on computer labs
  - Clarified training and scheduling requirements to department heads
  - Provided software and hardware for hands-on training in computer basics
  - Taught instructors procedures for teaching
  - WordPerfect — Excel — Lotus 1-2-3 — Microsoft Word

7. **Watch your grammar; avoid incorrect construction of statements.**

Faulty Parallel Statements	Improved Parallel Statements
Prepare scientific manuscripts for publication	Edit scientific manuscripts for publication
High skills in fund accounting	Prepare procedures for fund accounting
Excellent contract knowledge	Review contracts for errors

In the faulty category, the first phrase begins with a verb, the second is a noun phrase, and the third begins with an adjective. Stick with phrases that begin with strong verbs.

Avoid another grammatical trap: mixing noun and verb phrases in the same series.

Mixed Noun and Verb Phrases	Noun Phrases
1993–Worked as accounting assistant	1993–Accounting assistant
1991–Payroll Clerk	1991–Payroll Clerk
1990–Performed duties as bank teller	1990–Bank Teller

FINAL ASSEMBLY OF YOUR RESUME

Writing the Chronological Resume

Center your identification information on the page. Next add your power summary. Then transcribe the date/position employer information and the condensed paragraphs you wrote earlier in this chapter. Follow with each relevant category. Determine whether education comes at the beginning (new graduate or technical/scientist type) or at the end of the resume (for everyone else).

Finish the first draft, and check for omissions, irrelevancies, inaccuracies, and length. Make the necessary corrections. Type a second version, incorporating all changes. Study Chapter 5 to incorporate the visual impression you want to achieve. Give your resume for final critiquing to a trusted associate with a good command of spelling, punctuation, and grammar.

Writing the Linear Resume

Carefully review all the positioning setups in the section entitled "What the Linear Resume Should Look Like" in this chapter. With the sole change that all paragraphs will be single phrases set off by bullets or asterisks, the linear resume follows exactly the same steps for assembly as the chronological resume.

Writing the Functional Resume

Decide whether you will use paragraphs or linear style for your functional resume. (Remember that linear looks cleaner, but it takes up more space.) Put your identification at the top. Transcribe the functional headings and condensed paragraphs or linear phrases from the drafting forms earlier in this chapter. Follow that with a brief chronology of your work experience, unless you have a valid reason for omitting it. Next add education. Then review, correct, and condense. Type a second copy. Study Chapter 5 to incorporate the visual impression you want to achieve. Give your resume for final critiquing to a trusted associate with a good command of spelling, punctuation, and grammar.

Writing the Targeted Resume

Begin with your identification paragraph. Follow that on a separate line with your specific job target or objective. Use the phrase "In my targeted job, I can achieve the following:" or "Capabilities." Then list up to ten brief capabilities statements from your worksheets in this chapter. Follow that with "The following achievements directly relate to my job target:" or "Achievements." Below that list eight to ten accomplishments based on your worksheets in this chapter. Next comes the heading "Experience." Use no more than five lines to summarize your work history (dates, employer, title). If you have more positions, combine them. Follow this by "Education," using no more than two or three lines detailing school, degree or program, and dates. Make sure it fits on a single page. With extra space, add more accomplishments. Otherwise, cut until you have a single page. Study Chapter 5 to incorporate the visual impression you want to achieve. Give your resume for final critiquing to a trusted associate with a good command of spelling, punctuation, and grammar.

RESUME CHECKLIST

Now that your resume is written you should grade it by asking yourself the following questions. If you answer "No" to any of them, revise and correct.

1. Are there any typing or spelling errors?
2. Can the average reader understand what I'm saying?
3. Is my writing style simple, direct, and as powerful as possible?
4. Are paragraphs short and to the point?
5. Have I eliminated redundancies and similar examples?
6. Are my accomplishments clearly highlighted?
7. Do I cover all relevant time periods?
8. Is the layout simple, professional, and attractive?
9. Is this resume the best face I can show the world?

5

Polishing and Printing Your Resume

You've worked hard and your makeover resume is nearly complete. Don't let it fail on the basis of first impressions. Within three or four seconds your resume must pass the "Will I read more?" test. This is where balance and esthetics matter. Do you have enough white space? Does it have the proper elements in the right places? Is the typeface correct for the industry you're targeting? What about paper quality? Is everything spelled correctly? Are the headlines placed consistently? Is your use of attention-getting devices subtle but commanding?

MARGINS

In setting up your resume, you should have 1 to $1^1/_2$ inches of white space at the top and bottom of the page, and at least 1 inch on the left- and right-hand sides of the page. One reason is for visual effect; the page looks balanced. The other reason is practical—interviewers often write notes in margins. Give them the opportunity to do so, or they may unconsciously pass you by.

Your resume should also strike a balance from top to bottom. When you're done writing it, photocopy it and fold it in half lengthwise and then once again by width. The four quadrants of your resume as defined by the folds should be approximately equal. If your resume appears top or bottom heavy, or if the type tilts too much to the left or right, it needs editing.

CAPITAL LETTERS

Use caps sparingly and wisely. They *may* be used for your name. They should *not* be used for any other element in your heading. They are most effective in introducing the major sections of your resume: EXPERIENCE, EDUCATION, SKILLS.

UNDERLINING, ITALICS, BOLDFACE

Using computer word processing, once you have written your resume, you can go back and try out different effects. The key is consistency. If you underline your job title in your current position, you must do the same for each previous job. If you decide to have your resume typeset, you must specify exactly which segments require special treatment. Don't let the dramatic effects overpower the resume; you want them to high-light important material, not to upstage you. As stated previously, you must never underline by hand or use highlighter pens.

BULLETS, DASHES, ASTERISKS

Added special effects have their place in certain resumes. When writing a linear resume, you will use either asterisks (especially with typewriters) or bullets on separate lines to indicate each key piece of information under a major heading.

ACHIEVEMENTS

- Managed business relationships with top-level financial executives
- Supervised installation of electronic quotation equipment for 100 branch offices

OR

Assistant Manager Technology Council   1987-Present

- Managed publication and distribution for <u>Advanced Technology News</u> (layout, printing, mailing)
- Researched market for special seminar for institutional investors

---

INDENTING

Separate various kinds of information by indenting. The computer tab key is handy for this. Use two, or at most, three degrees of indenting to make the reader's eye follow the logical relationship of information.

EDUCATION

Westminster College, Albany, NY
B.S. Chemistry
Economics Minor
>   Financed expenses by related part-time work, grants, and partial
>   university scholarships.

HEADERS

Despite your best efforts, or because your accomplishments are so lengthy, your resume may run to two pages. If it does, use a header at the top left of page two with a phrase like "Page 2, Resume—George Stone." Be sure the page break occurs at the end of one paragraph and the second page begins at the start of another.

EDITING TO TIGHTEN SENTENCES

- **Use short sentences.** Don't use five words when three will do! Not: "I edited the college yearbook," but "Edited college yearbook."
- **Avoid repetition.** If you performed the same payroll function as clerk *and* assistant to the head of the department, list it once. Give one example, not three.
- **Use a positive tone, with dynamic words.** Not "Changed the filing system around," but "Streamlined filing system."
- **List only your most recent jobs.** I've said it before, I'll say it again: Keep your earlier experience as deep background. Condense it like this:

  "1972–1983 A variety of secretarial and word-processing positions."

MAKE ONE FINAL CHECK OF ALL KEY POINTS

You're probably sick of reading and rereading your resume by now. Do it once more to catch any last-minute errors, before finish-typing, word processing, or typesetting. What should you look for?

- Your resume fits neatly on one page; two if it's absolutely critical.
- There are no spelling, grammar, or punctuation errors.
- Your personal identification block is centered at the top.
- Margins are 1 to $1^1/_2$ inches at the top and bottom, and 1 inch on each side.

- The layout is easy to read; indentations are logical.
- No paragraphs are longer than eight to ten lines.
- You have prioritized importance of material using proper highlighting techniques like underlining, and so forth.
- Action verbs communicate results.
- All extraneous personal information has been deleted.
- You've eliminated all redundancies.
- The overall resume has visual appeal, compelling the reader to give it serious consideration.

TURNING YOUR FINAL DRAFT INTO TYPE

In a decade, the computer went from being a luxury to a virtual necessity. But you may not own one or perhaps lack knowledge of word-processing programs. Can you use a typewritten manuscript for your final copy? That depends on who you are.

If you're a college senior, just starting out in your career, or looking for nonprofessional work in offices or other environments, a typewritten resume is acceptable. A midcareer executive resume is expected to look polished—whether by word processing or typesetting. And for a senior executive, a typeset resume is the best approach.

WHICH TYPEFACE IS BEST?

Choice of typeface depends on what kind of position you're seeking. For a typewritten resume, the most effective typeface is Courier. The typing should be professional and error-free. Your resume should be typed on an electric or electronic typewriter (even printed on a letter-quality computer printer) with a carbon, not a cloth, ribbon. No erasures, whiteouts, or smudges.

For computer word-processing programs, it depends on which you're using. Some fairly standard typecases include Courier and Times New Roman. Your basic text should be no smaller than 12-point, or 10 if you're really crunched for space.

If you have your resume typeset, 10-point is the smallest to use. The more conservative typefaces are Times Roman, Century Schoolbook, and Palatino. They are readable, available, and acceptable. They are all serif types, which are considered to be traditional and businesslike. If your activity allows more latitude of expression (graphic artist, designer, architect, performing artist, etc.), Optima is an example of a readable, contemporary sans-serif type.

SELECTING THE PROPER PAPER

Need I say you should use $8^1/2$" x 11" paper, the standard we all know and accept? I met one eager candidate so determined that his resume should fit on one page that he created an $8^1/2$" x 14" version. There are cases, as I said, where it is permissible to go to two pages. Just make sure it's two $8^1/2$" x 11" pages! And never have your resume printed on both sides.

Use black ink only, on quality white or ivory paper. Gray is acceptable, but sometimes it's hard to photocopy. Shop around for paper; today's office warehouse stores usually have a good variety. If not, ask for samples from your printer. What you want is at least 24-pound bond, preferably with a cotton rag content. While you're buying paper, also get matching blank stock to use for the cover letter (see Chapter 7) and matching envelopes.

6

Writing the Cover Letter

Whole books have been written about letters used in job searches. Basically, they fall into two categories—letters that accompany resumes and those meant to stand on their own. In any case, the letter's purpose is to personalize your achievements—to make you stand out from the pack. The well-written letter should be a response generator, leading the reader to review the resume or call you.

RESUME ENCLOSED: RESPONDING TO WANT ADS

I'm not talking about blind ads with post office boxes. Forget them. When you respond to a worthwhile-sounding position based on a want ad from an identified employer, be aware that a juicy opening can generate hundreds of responses. Realistically you can't expect a lot from such long odds. So what can you do to strengthen your case?

- **Use the correct letter-writing format.** Begin with a date centered or flush left. Skip two lines. Follow with the name of the individual, his or her title, name of firm, and address. Skip two more lines. Identify the specific position you are seeking, based on the ad. (The company may have more than one job listed. If you prefer, you may mention the place where you saw the ad in the lead paragraph.) Skip an additional two lines and use the correct salutation: Dear Mr. or Ms. followed by the last name, spelled correctly. (Verify the spelling.) If you don't have a name, begin with the body of the letter. Don't write "Dear Sir," or "To whom it may concern."

REPRODUCING YOUR RESUME

With quick-printers around every corner, you should not have difficulty finding one who will reproduce your resume by photo-offset. In essence, a picture is taken of your finished resume, and the results are printed on your high-quality paper. This is the preferred method for best-quality results. Because of high competition, prices are now quite reasonable.

What about photocopying? It's a controversial subject. Even if you use good-quality paper, there are times when the photocopier has specks on it that get reproduced right onto your finished product. If you can't avoid photocopying, make sure you get a sample first.

When you have created your resume using a computer word-processing program and need a small quantity to send out in a hurry, you *may* use your computer printer, if it's a high-quality laser model or if your dot matrix printer is exceptionally clean. With technology advancing so fast, better-quality computer printers are ideal in certain cases, such as for targeted resumes, where you're changing focus for several different versions.

THE NEXT STEP

In Chapter 6, you'll learn how to create a cover letter. Although it is not always necessary, there are times when it is indispensable.

- **Pack as much punch as possible into the lead paragraph.** There are several ways to do this. One is by putting your leading qualification into one or two succinct sentences:

  With over seven years' experience as an executive secretary, I am a strong candidate for the position you advertised in today's <u>Las Vegas Review-Journal.</u>

  OR

  As a Loan Officer at First National Back, I originated more loans than any other employee in the past five years. That's why I think we should discuss the position you advertised in today's <u>New York Times</u>.

  Another way that works for creative fields especially is to show genuine enthusiasm for the perceived opportunity:

  What a wonderful position for someone with an art and marketing background! Your ad in today's <u>Detroit Sun</u> for a self-starter to market your art leasing programs caught my eye and sent me to my computer immediately. The job appears perfectly tailored to my background, interests, and experience—almost as if I had written the qualifications for myself.

- **Follow your opening paragraph with a brief summary of qualifications.** I emphasize the word *brief*. You don't need to repeat what is already stated in your resume. Just pick out the highlights. There are two ways to do this: One is by placing them in the second paragraph. The other is to divide the page in half. One the left side, you put the advertised qualifications; on the right side, you list your own. (See samples on page 82.)

- **State any other considerations that drew you to the ad, such as geographical reasons.** (Optional)

- **Indicate any special rationale for making such a move**—not "I was recently let go" or " I see this as a better opportunity," but if there are "special circumstances" that might reflect favorably on your candidacy, be sure to indicate them:

  I have several years' management experience with The Golden Grainery. Unfortunately, the owners have announced their retirement and are closing the store within three months. Your ad could not have appeared at a better time.

  Other satisfactory reasons may include massive layoffs and economic shrinkage in your geographic region.

- **What about salary?** I don't favor mentioning it in your resume. In a letter, it's a more personalized thing. If a range is stated in the ad, obviously it meets your criteria or you wouldn't be responding. If no salary range is mentioned and they ask that you reveal your bottom line, personally I would simply not mention it.

- **Contact information (where and how you may be reached), request for an interview, and a "Thank you for your consideration"** should close out the letter.

REGRETTABLE RESPONSE LETTER TO WANT AD

[Problem: Too General]

February 7, 1994

F&Z Confidential Reply Service
Dept. 3525
7242 Orangethorpe Avenue
Buena Park, CA

To whom it may concern:

I am forwarding you a copy of my resume in regards to your advertisement in the <u>Los Angeles Times</u>, February 7, 1994 for a Security Director.

As you can see my management experience is very extensive. I feel with my background and my commitment to excellence, I would be an asset to your growing and challenging corporation.

I look forward to meeting and discussing your opportunities.

Yours sincerely,

Cyrus Quincy

Stanley Darden
5783 Winnetka Avenue, #54
Mission Hills, CA 93555
(818) 939-4494

May 23, 1993

John Sylvan
Professional Employment Department
Octagon Hotel
444 N. Figueroa Street
Los Angeles, CA 90005

Dear Mr. Sylvan:

I have enclosed my resume in response to your advertisement for a Director of House-keeping that appeared this week in the Los Angeles Times.

I have numerous years of successful management, supervisory, and hands-on experience in the environmental and custodial services area, with special emphasis on floor care. I also possess an in-depth knowledge of the various chemicals used in janitorial and custodial duties, and have established policies and procedures still being used by major medical facilities and industrial complexes in the Greater Los Angeles area.

You will find me to be self-motivated, energetic, and dedicated to a job well done.

I look forward to meeting with you in person so that I may better demonstrate my abilities.

Thank you for your interest and the courtesy of your response.

Sincerely,

Stanley Darden

Enclosure

RESUME ENCLOSED: BROADCAST JOB SEARCH LETTERS

If you plan to devote a substantial portion of your job search efforts to a direct mail campaign, you will use a variation of the broadcast letter in several circumstances:

1. Referrals—Where you have a specific contact; this is your best chance to land an interview.

2. Name-based query—Based on research using trade directories, professional association rosters, electronic databases, or other industry sources, you may acquire the name of the individual who heads the department relevant to your job target; for example, for a marketing professional, the marketing director at Avis Rent-A-Car, Southest Division, or for a secondary school teacher, the principal of several nearby high schools. Do *not* send to Human Resources; it's a waste of time.

3. "The big broadcast"—No specific name of a person; addressed by title to a department head, like "Chief Financial Officer" or "National Sales Director." This represents the widest and least promising casting of your employment fishing net. It's so broad, that even if you send out 100 letters, you may consider yourself very lucky to get one or two responses. Frankly, I wouldn't waste my time with this approach, so I'll limit discussion and examples to the first two.

Elements of both letters are similar to the format in the first section in this chapter, "Responding to Want Ads." They include: (a) Your name, address, and phone number; (b) Your contact's name, his or her position, employer, address, etc.; (c) Salutation; (d) *Strong* opening paragraph; (e) Statement of purpose; (f) *Brief* summary of qualifications; (g) Optional statement about reason for making change; (h) Optional geographic considerations; (i) Response request.

SIMON W. CONTRERAS
9 Forest Gate
Simi Valley, CA 92679
(805) 882-3697

May 23, 1994

John Sylvan
Professional Employment Department
Octagon Hotel
444 N. Figueroa Street
Los Angeles, Ca 90005

Dear Mr. Sylvan:

In response to your ad in the Sunday, May 20, 1994 edition of the *Los Angeles Times* for a Director of Housekeeping, I feel my qualifications favorably match your requirements.

YOUR REQUIREMENTS	MY QUALIFICATIONS
• Solid experience in property maintenance	• 15+ years' experience and physical plant maintenance management including:
• Minimum five years housekeeping	• 7 years housekeeping supervision
• Optional grounds management background	• Grounds management experience
	• Custodial operations
	• Security systems
	• Fire prevention

Enclosed, please find my resume for your confidential review. Salary history may be discussed during an interview. I look forward to meeting with you in the near future.

Sincerely yours,

Simon W. Contreras
Enclosure

REGRETTABLE REFERRAL LETTER

[Problems: Insufficient specificity regarding experience. Lack of telephone number on letterhead. No firm request for appointment.]

2326 Spanish Trails
Tampa, FL 33262

February 23, 1992

Amanda Rogers
Co-Owner
Andover & Rogers
43 Market Place Lane
Coral Gables, FL 33632

Dear Ms. Rogers:

Warren Bridges at the Coral Gables Country Club suggested I contact you about the opening in your bookkeeping department. I have worked for Warren over the last four summers as a clerk and special assistant during the busy vacation season.

As you can see from my resume, I will be graduating this semester, with a degree in business and some related bookkeeping experience.

Yours is the kind of opportunity I have been looking for. I live nearby and could come for an interview at your convenience.

Thanks for your consideration.

Peter Newbridge

APPROPRIATE REFERRAL COVER LETTER

Stephanie McDaniels
456 Stapleton Road, #26
New Haven, CT 03111
(305) 888-4442

January 10, 1994

Jason Richman
Editor-In-Chief
APPLICATIONS IN SCIENTIFIC GRAPHICS
8550 Farmingdon Road
Bridgeport, CT 03853

Dear Mr. Richman:

Stephen Young, your former roommate at Yale, suggested that I contact you about a possible position as a technical editor. He said that when you met for lunch recently, you had mentioned the possibility of a vacancy arising quite soon.

I welcome the opportunity to present my resume, which summarizes six years of experience in the technical writing/editing field. As the technical editor for a small scientific journal, I did all the editing as well as some writing. Most recently, I was in charge of supervising two assistant editors for a large scientific journal. I am especially adept at translating scientific verbiage into lay language.

Thank you for considering me for this position should it become vacant. I will call you next Wednesday concerning my interest in the position.

Sincerely,

Frances McDowell

REGRETTABLE NAME-BASED QUERY BROADCAST LETTER

[Problem: Too wordy. Doesn't make his point quickly. This senior candidate should know better!]

George Carlisle
5312 South Jennifer Way
Englewood, CO 80112
Phone: (393) 923-4530

March 6, 1994

John McGill
President
STARLINE FASHIONS, LTD.
824 Park Avenue
New York, NY 10034

Dear Mr. McGill:

Time enough is still a problem, isn't it? Not scarcity of printout data, but time.

Perhaps you need more time to evaluate the conflict in information you already have. The conflict that you built in because you delegated profit responsibility to people, not to computers.

Has a Corporate or Division President anything more important to do than taking the time for good decision making? Can some of the things he does, or wants to do, be handled by an Assistant having similar experience, maturity, and profit orientation? <u>This is what I offer:</u>

* Comprehension of the essential elements in your business with the ability to establish, improve, or implement programs and procedures that contribute to corporate performance.

* Experience in saving substantial sums through operational and market analysis and controls.

I send my resume in confidence, since my present company is unaware of my plans. May I talk with you soon?

Yours truly,

George Carlisle
Enclosure

RUTH BENJAMIN
8437 Wyandotte Lane
Bellevue, WA 98128
(206) 449-3579

October 7, 1993

Harry Washburn
Senior Vice President
The Groverton Fund
92 Wall Street
New York, NY 10016

Dear Mr. Washburn:

Will you take 30 seconds to read this letter? I'm looking for a new challenge to apply my substantial knowledge in Fixed Income Portfolio Management, and I'd like that challenge to be The Groverton Fund.

As Fixed Income Portfolios Manager of a leading investment counsel firm in Bellevue, I have full responsibility in an area of strong growth. I am experienced in aggressive bond management for large pension funds, insurance companies, banks, and others with sizable portfolios.

I am agreeable to a reasonable amount of travel and would relocate to New York if the potential were right. Earnings are negotiable above total current income.

My resume is enclosed in greatest confidence. Due to the time differential, I will call you from my home at the start of your workday next Tuesday.

Sincerely yours,

Ruth Benjamin

Enclosure

RESUME ENCLOSED: LETTER TO EXECUTIVE SEARCH FIRMS

You may also use this letter format to send your resume to an employment agency. What are the differences between the two? The executive search firm (headhunter) is retained by the client to produce candidates for client companies. They do *not* work on your behalf. If you have the proper qualifications, though, they're only too happy to match you with one of their clients. If your credentials are sound, but no current need exists, chances are you'll get entered into their database. Either way, you get most benefit from headhunters when you have an established, successful employment track record.

Employment agencies can and do work for either the client firm or you, the employment candidate.

When writing to an executive search firm, do *not* ask them to help you find a job. It makes you look unsophisticated and unaware of how they work. Instead, you would begin by referring to any client searches they might be conducting that would fit your background.

Many elements of the cover letter to executive search/employment agencies parallel other "resume enclosed" letters already discussed. These include: (1) Your identifying information; (2) Search firm or employment agency identifying information; (3) Salutation; (4) Introductory paragraph or statement; (5) Your job objective; (6) Brief summary of qualifications; (7) Reason for making change (you can be a bit more candid than you would be with a prospective employer); (8) Salary qualifications (this is one time that it might be appropriate for you to disclose this information; no point in having your resume go to prospective employers who do not fit your economic criteria); (9) Geographic or other preferences (you don't want to limit yourself too much, but you do need to explain your desire to move or any restrictions in your candidacy); (10) Statement of willingness to provide additional information; (11) Contact instructions for reaching you; (12) Thank You.

REGRETTABLE EXECUTIVE SEARCH FIRM QUERY LETTER

[Problem: Too general; too generic.]

Philip Wanstadt
3 Keystone Avenue
Skokie, IL 60204
Tel: (608) 478-0031

January 6, 1994

Hazelton & Overby
Executive Searches
548 Michigan Avenue
Chicago, IL 60609

Gentlemen:

I am writing to you in hopes that you are currently providing service to client firms seeking uniquely talented sales and marketing management professionals. With a consistent track record of success in the plastics and specialty chemicals industries, I believe that I could contribute immediately to a variety of business development situations.

I have enclosed my resume to provide you with details of my background and skills. I would be most anxious to discuss my career goals with you - even if only on a purely exploratory basis.

If you have any immediate questions, do not hesitate to call. Should you have employment opportunities that you feel may be of interest to me, I would appreciate hearing from you.

Thank you for your consideration, and I look forward to hearing from you.

Yours sincerely,

Philip Wanstadt

Enc.

Randal Cunningham
934 Sunnyslope Drive
Dallas, TX 76424
(214) 934-6783

February 8, 1993

Alice Walker
Scoresby, Smith, Simonton
809 Peachtree Plaza
Atlanta, GA 33481

Dear Ms. Walker:

I understand that one of your firm's specialties is executive searches for companies who need drastic turnaround.

For almost 10 years I have been engaged in the task of taking over the management of companies or divisions that were in trouble or heading that way. In four specific instances I have been responsible for major improvements.

At this time, the wholly owned subsidiary corporation that I manage is in the final stages of being sold, and I doubt that there now exists a spot in the parent company for someone with my entrepreneurial skills.

Therefore, I am looking on two levels: 1) Finding a company with steady growth potential in which I can take a long-term equity position (in this case, compensation levels are secondary to growth potential). 2) Moving into another "turnaround" situation with compensation at least 15% above current earnings.

Though these might appear to be disparate goals, either is acceptable based on the criteria explained above. I am geographically flexible.

My resume gives a brief summary of my total qualifications. I would be pleased to expand on it personally at your convenience should you have an appropriate opportunity with one of your clients.

Thank you for your consideration.

Randall Cunningham
Enc.

RESUME LETTERS: IN PLACE OF THE RESUME

Why would you want to use a standalone resume letter, instead of the traditional resume–cover letter combination? For three basic reasons: (1) The resume letter can be customized to the needs of a specific employer, even beyond that of the targeted resume. (2) It can be used in almost any situation—whether you're responding to a want ad, following up on a referral, or even in a broadbased job search. (3) In certain cases it's more effective, especially if you're either at the pinnacle or at the beginning of your career. That's when I suggest you try it.

On the other hand, there are certain disadvantages. It doesn't look like a resume, so it may be a problem if you're seeking employment with more traditional or large firms.

When using a resume letter, do *not* send it to the human resources department. Be sure to include a follow-up statement in the letter. Your phone call can make the difference in actually securing an interview.

WHY A RESUME LETTER FOR THE SENIOR EXECUTIVE?

If you're an executive with a salary in the $50K range or above, you would want to address your resume letter to the president (of a smaller company) or appropriate executive vice president of the company. The letter should present the highlights of your background and experience. It should include a description of your major achievements, plus the reasons you want to work for that specific firm. Do your homework regarding that company's background, strengths, and needs before writing. Your letter should *not* include a listing of jobs present and past—that is the exclusive function of your resume.

MICHAEL GROVES
211 Coyote Lakes Avenue
Wilmette, IL 60018
(312) 434-2299

March 23, 1992

Kenneth Moreland
Vice President/Production
HARLAND STEEL
3527 Overby Road
Cleveland, OH 43216

Dear Mr. Moreland:

A few months ago I read in <u>The Wall Street Journal</u> that your firm is part of the sizable contingency that will be entering contract negotiations next fall. Given the recent rebound in U.S. steel production, you must be anxious to keep things moving smoothly. This past week I saw your ad for a director of labor relations in the <u>National Business Employment Weekly</u>. That's where I feel I could assist you. Given the growth picture at Harland, I would welcome becoming part of your labor management team.

I have over 20 years experience in labor relations, including the last 7 as assistant manager for labor relations at Benninger Steel. My accomplishments include:

- Negotiating a comprehensive local seniority agreement with 2000 member local product and maintenance union.

- Averting wildcat strikes threatened in 1990 with three of the largest local unions in the steel industry.

- Counseling operating management in all phases of labor relations including suspension and discharge; all problems satisfactorily resolved at plant level.

I will be in Cleveland in two weeks, and would appreciate the opportunity to meet with you. I will call next Tuesday to see if we can arrange a mutually convenient time. Should you want to see my complete resume, I will be pleased to forward it.

Yours sincerely,

Michael Groves

Peter Hanson

Permanent Address:
5310 West 65th Street
Waco, TX 78712
(512) 827-3190

Temporary Address:
4532 Excellon Avenue
Austin, TX 78717
(512) 734-5777

February 26, 1994

John McSweeney
Manager/Chemical Additives Department
MOBIL OIL CO.
4170 Torrance Blvd.
Torrance, CA 90512

Dear Mr. McSweeney:

When Mobil's representative, Henry Folsom, was on campus recently at our Job Fair, he suggested I contact you close to graduation. That time is here. I plan to move to California and am very interested in an entry-level position in petroleum additives.

I am a Baccalaureate Degree candidate in Petroleum Engineering, and anticipate graduating in May. I am in the top two percent of my class and have maintained a 3.8 average on the 4 point scale.

During my four years at the University of Texas my related coursework has included Oil Well Drilling, Drilling Design and Production, Petroleum Engineering Design, Well Treating and Evaluation, Rock and Fluids Lab, and Reservoir Engineering.

I have been president of the Society of Petroleum Engineers on campus for 1 1/2 years, and secretary of the Engineering Club for six months during my sophomore year.

In the course of my last two school years, I have worked in the Instrumentation Lab, assisting in the development of a Phase Compactor system for the study of plasma physics. I also designed an active bandpass filter circuit with predictable phase response.

I'll call next week for an appointment during our Spring break, March 12–19.

Yours sincerely,

Peter Hanson

COVER LETTERS FROM START TO FINISH OF YOUR JOB SEARCH

While I emphasize using your resume or resume alternative to generate job leads, there are other reasons for writing letters to further your search. Even before you begin trying to schedule interviews, you can try a tactic known as the *information approach.*

In this letter, your purpose is different. You are not trying to get an interview (well, sure you are, but you're doing it differently). In this letter, you contact a respected senior person in your field of interest and ask if you can seek information, advice, and referrals. This kind of letter usually achieves two vastly different responses.

1. He or she may see through the ploy (particularly if you are well along in your career), and choose not to see you.

2. The person is pleased that you think of him or her as someone of mentor capabilities whose opinion is valued, and may take the time to see you. If you do get the opportunity, be prompt and take copious notes. Otherwise, it will look like your sole purpose is to use him or her as an entree into the firm.

Variations of this letter depend on whether it is being sent based on a mutual acquaintance, or strictly cold turkey. The stronger the source of the referral, the better your chance of landing the meeting.

Following such a meeting, or a job interview, you'll be well-remembered with a polite thank-you note. Most people don't bother, but it could make a difference.

REFERRAL INFORMATION APPROACH LETTER

Wanda Winston
14 Allenford Court
Raleigh, NC 27853
(413) 669-4327

July 3, 1993

Sandra Fishburn
Partner
Connors, Jamison, Fishburn
Attorneys At Law
8344 N.W. Avenue 26
Washington, DC 20094

Dear Ms. Fishburn:

As you probably know, your college associate at North Carolina State, Harriet Johnson, is now teaching in our law school. When I spoke to her concerning my interests in contacting firms specializing in environmental law, she suggested I speak with you.

This is my background: While completing my law degree at North Carolina State, I have become extremely interested in environmental law, especially as it relates to water pollution. Many of the cases I reviewed involved law firms, lobbyists, nonprofit environmental groups, and federal agencies in Washington. Therefore, with degree in hand, it is logical that I plan to pursue my career in our nation's capital. I have been active in the North Carolina State Law Society including such positions as vice president and treasurer. I graduated in the top 5% of our class of 210.

I know I could learn much from your experience and would appreciate any information, advice, or referrals you might give me. I will call you next Wednesday to see if your schedule permits you to meet with me. Thanking you in advance, I am

Yours sincerely,

Wanda Winston

COLD TURKEY INFORMATION APPROACH LETTER

Ralph Warren
31 Mentone Avenue
Brisbane, CA 91423
(415) 888-3201

March 4, 1992

Jacob Cavendish
Vice President/Sales
Warren Publishing
5341 Old Washburn Road
Santa Barbara, CA 92452

Dear Mr. Cavendish:

I'm sure that much of the credit for your firm's 28% jump in space sales this past year is due to the highly motivated staff you recruited during the past three years. I hope to work in such a position for a company just as committed to growth.

I have been financing my four years at Santa Clara State University through on-campus magazine sales through the *Reader's Digest*. It has given me a solid feel for the sales end of the publishing business and I want to pursue a career in advertising sales.

In beginning my job search, I am trying to gather substantive advice and direction from those in the field. Accordingly, I hope you will have the time to see me within the next month and give me the benefit of your expertise.

I'll call on Monday, March 10, to see if your schedule permits you to see me. I look forward to meeting you.

Yours sincerely,

Ralph Warren

Monica Wenders
443 West 6th Street
New Orleans, LA 54313
(372) 866-3107

January 14, 1990

Gerald Grantham
Marketing Manager
Studio Associates
23 Riverfront Walk
New Orleans, LA 54322

Dear Mr. Grantham:

Sally Walker was right when she said you would be most helpful in advising me on a career in marketing.

I appreciated your taking the time from your busy schedule to meet with me. Your advice was very useful and I have incorporated your suggestions into my resume. I'll send you a copy next week.

Again, thanks so much for your assistance. As you suggested, I will contact Marvin Watkins at Glenby & Wilson next week concerning a possible opening with his firm.

Sincerely yours,

Monica Wenders

Jason Richards
589 West Mannington Blvd.
Portland, OR 98423
(509) 490-6753

May 23, 1992

Glenda Ryerson
Human Resources Director
City of Morganville
8953 Old Road
Morganville, CA 92311

Dear Ms. Ryerson:

Many thanks for the opportunity to interview with you and your assistant, Henry Silvers. After our discussion, I have great confidence that I would be a valuable addition in the capacity of Director of Parks and Recreation.

With my background in all phases of community recreational programs during my three years here in Portland, I am sure I could bring the kind of innovation and leadership you feel is essential to the success of the Morganville revitalization program. I am also quite certain that I could substantially expand the programs without a budgetary increase. This is an important consideration given California's lean economy.

Please let me know if I can provide any further information that will help you in your hiring decision. I shall check with you next Tuesday to see how your selection process is going.

Yours sincerely,

Jason Richards

7

Electronic Resumes

You just can't escape technology. Even the most creative types have a computer link somewhere. The drafter has his or her CAD/CAM system. The advertising designer draws, selects type fonts, scales ads, and creates finished art on his or her friendly Mac. Even the hair stylist would feel adrift without his or her computerized appointments and bookkeeping systems.

That being the case you'd figure that computerized technology would invade the world of human resources. If your job search is going to be tied to major corporations, online modem-based services, and/or headhunters, you need one more resume variation.

HOW TO LABEL YOURSELF

Don't throw up your hands and run screaming—it's not that difficult to modify your finished resume. Basically, what you're going to do is create a set of labels for yourself; the more labels the better. These labels (called *keywords*) are scanned by computerized systems to see if you meet the parameters of specific jobs.

You know how I stressed using active verbs to begin the phrases that make up your resume. They're great for the human eye. But the computer wants nouns, and that's what keywords are. They *name* your skills, talents, experience, and abilities.

In their book *Electronic Resume Revolution* (Wiley, 1994), California career-columnist Joyce Lain Kennedy and business editor Thomas J. Morrow offer up multiple resumes in this new configuration. What's different is primarily what's on top. Here's two samples, one for a teacher, the other for a purchasing manager.

Terrance A. Thompson

201 Larkspur Lane
Pleasant Valley, NY 15000
(318) 675-1212

KEYWORDS: Child Care. Teacher. Early Childhood Development. Mathematics Teacher. Private School. Day Care Center. Primary Age Students. Natural Science. Reading. Drawing. Elementary Education. MA Degree. BA Degree. San Diego State University. University of New Mexico Doctoral Student.

OBJECTIVE: Position in child care or teaching at a university observational school or day care center—preferably at a school where I can complete my doctorate in Early Childhood Education.

SUMMARY OF QUALIFICATIONS: Successful with the challenge of teaching groups of children. Patient, confident, and committed in working with children. Teaching credentials in three states: New York, New Jersey, and Pennsylvania.

[Thompson then goes on to list his employment history, education, and credentials in a modified targeted resume format.]

Stephen J. Sterling
19917 Keeamoku Street
Honolulu, HI 98000
(808) 721-7211

KEYWORDS: Purchasing Manager. Purchasing Supervisor, Purchasing Agent. Industrial Buyer. Supply Manager, CPM, NAPM, Electronic Components. Raw Materials, Material Rejection Board. RFQ.

[Sterling's resume continues in classic chronological fashion listing experience, education, and other qualifications.]

HOW DO YOU WRITE A KEYWORD SUMMARY?

Kennedy and Morrow advise using the *inverted pyramid* style—the story format favored by journalists. List the most important parts of your story first:

> In your keyword summary, get to the more important stuff immediately. The usual order is to list the job title, occupation, or career field you want. Next come your most important sales points, especially the essential skills for a specific position. End with your education information.
>
> As a new graduate with little experience, you may or may not want to alter the lineup, reversing education and experience. Your choice depends on which assets are your most salable.

Kennedy and Morrow further advise resume writers to avoid continually using the same words in the body of their resumes. Use synonyms, they suggest. For instance, if your keyword summary contains the term "news anchor," use "TV journalist" in the body of your resume. If "RN" is in your summary, write "Registered Nurse" in the body.

> No matter how you construct your keyword resume you must make it easy for a human reader to see the direct relationship of its components. Your resume must hang together with meticulous internal logic. Its body must provide a match with your keyword summary.

OTHER MAJOR DIFFERENCES IN ELECTRONIC RESUMES

- One page is not necessarily the rule here. For new graduates, yes; most people can go from one to two pages. Senior executives have the license to run two to three pages. That's because it's the computer doing the scanning, not some bleary-eyed human resources person.
- Cover letters are desired. Recruiters say they want them to amplify the resume and that they are electronically stored along with the related resumes.
- Use simple type that won't "muddy up" when it's read by scanners and Optical Character Readers (OCRs). Sans serif (without the little flourishes at the ends of each letter) is preferred to serif type faces. Keep the resume strictly no frills. To accent key phrases you may use boldface or capitals. Avoid italic text, script, and underlined passages. Minimize the use of general abbreviations; maximize the use of industry jargon and abbreviations.
- Coordinate your resume to reflect a specific concept. Kennedy and Morrow explain, "The function of marketing is a concept; trade shows, marketing research, and focus groups are the examples. Human re-

sources is at the concept level and employee benefits, 401K plans, and compensation analysis are at the amplification level."

- Use only the last 10 to 15 years of your worklife. If you're light on experience, use only years instead of months and years.

- Keep track of every employer who gets your electronic resume. Then update it every six months.

- Don't send multiple resumes to different parts of a major corporation. The computer scanning devices will usually pull up everyone with the same name and telephone number. Multiple resumes tend to make a recruiter think, "This person doesn't know what he or she wants to do."

- Do list any professional societies you have joined. They have status in the electronic resume world. Why? Because they're viewed as highly searchable skills (keywords), a recruiter will hone in on them when doing special recruiting.

- Follow up after submitting your resume. Kennedy and Morrow advise: "Learn the name of the automated tracking system's administrator or operator/verifier. Ask that person, 'Did you receive my resume? Was I a match anywhere? Has my resume been routed? To whom? Which department?' If you are stonewalled, or asked why you want to know, simply tell the truth: 'I need to follow up on my routing.'"

- Use plenty of white space—computers like it. According to Kennedy and Morrow, "They use it to recognize that one topic has ended and another has begun. A scanning industry maxim is 'White space makes errors go away.'"

- Use common language that everyone knows. Kennedy and Morrow say "Don't describe a skill as 'Can pound iron spikes.' Instead, say, 'Can hammer nails.' Give your resume the 'common touch' test."

- Once hired, use applicant tracking systems as a career management tool. Provide a yearly update of your resume to every database you hope will be a steppingstone on your career path.

- After you're employed, your resume moves from applicant tracking to employee tracking. It's in there working for you should advancements that mesh with your skills appear in the company. Kennedy and Morrow suggest, "Make friends with the system administrator and operator/verifier. You want to be sure your updates are entered promptly, and there's no better way than to be on good terms with the humans who get those machines moving."

Using the computer as a tool that can maximize your exposure is the best way to view this new twist on resume writing. Once you master the art of writing the electronic resume, you'll have an important edge over the 95 percent who are not taking advantage of technology to tap into every available niche where the right job may be hiding.

8

Resume Samples

PROBLEM RESUME: ACCOUNTANT

Andrew M. Erikson
25 Hilltop Road
Birmingham, AL 35204

① CAREER OBJECTIVE: To join a major corporation as accountant and eventually become controller or treasurer.

Education

② Bachelor of Science, Accounting; University of Alabama, 1983-1989
Attended college evenings while working full time in controller's department of Gulfstream Enterprises and Davis Aerospace Corporation.

Advanced Accounting Courses | Business Background Courses

Managerial Accounting
Accounting for Decision Making
COBOL I & II
Seminar in Management Accounting

Financial Management
Industrial Economics
Management Information Systems
Strategic Management

③ Experience
Staff Accountant, Gulfstream Enterprises, Huntsville, Alabama
1988-Present

Worked on various assignments in the Controller's office using IBM ledger system. This related to computer accounting training at the University of Alabama and enabled me to accept increased

④ responsibility. Desire change because my degree qualifies me for advanced responsibility not possible in present position. Discussed this decision thoroughly with the Controller, who requested that I remain with Gulfstream until I find a new position.

Accounting Supervisor, Davis Aerospace Corporation, Birmingham, Alabama   1983-1988

Full-charge and general-ledger bookkeeping. Worked with IBM ledger system, prepared financial statements, and rotated as backup accounting clerk as needed.

⑤ References
Edward N. Albrough, Controller
Gulfstream Enterprises
34 River Avenue
Birmingham Alabama 35204
(205) 999-4567

Howard Stone, Ph.D.
Professor of Accounting
University of Alabama
P.O. Box 345
Tuscaloosa, Alabama 35401
(205) 654-3219

General Issues

Language of this resume is weak. The sentences are too long, the vocabulary too loose. Erikson conveys low self-esteem. Otherwise, I think he would have presented himself in a more positive, persuasive manner. It's too bad, because he did well in his last two jobs, and in college, too. What he needs is more punch, something he can achieve with action verbs and a linear format.

1. Omit objective. Convert to power-packed summary of achievements.
2. Move education to end of resume.
3. Use strong action words to describe accomplishments and effectiveness.
4. Omit reason for leaving previous jobs.
5. Omit names of references.

Andrew M. Erikson
25 Hilltop Road
Birmingham, AL 35204
(205) 776-8975

Summary

Seven years as accounting supervisor and accountant with increasing responsibility and proven performance in a sophisticated computer-based accounting environment. Advanced training and successful experience in computerized accounting systems.

Experience and Accomplishments

1988-present   **Gulfstream Enterprises, Birmingham, Alabama**

Staff accountant reporting to controller of $20-million oil exploration company.
- Introduced, installed, and monitored new IBM system.
- Trained accounting staff of 12 in its use.
- Identified significant error problem and instituted new system of financial reporting to correct it.
- Efforts resulted in a reduction of manpower for basic accounting function and preparation of financial reports, with annual savings of $40,000.

1983-1988   **Davis Aerospace, Inc., Birmingham, Alabama**

Accounting supervisor in audit department of $50-million aerospace contractor. Received increasing responsibility with exceptional results while studying evenings for degree.
- Mastered full-charge and general-ledger bookkeeping duties leading to supervision of six bookkeepers within one year.
- Instituted time- and cost-saving accounting procedures.
- Reported directly to Auditor as company's chief computer systems troubleshooter.

Special Skills

COBOL I and II programming and systems analysis

Education

B.S. Accounting, University of Alabama, *magna cum laude*

References

Provided upon request, once mutual interest has been established.

SONYA McREADY

(1) OBJECTIVE

Copywriting position in which I can use my expertise in food marketing.

(2) WORK HISTORY

1987-Present   **(3)** <u>Senior Copywriter, Burton Advertising</u>. Chief copywriter on five major food accounts, responsible for two monthly magazines published by clients, one newsletter, and several miscellaneous reports and booklets geared to the consumer market. Wrote a cookbook for a major client in 1989.

1982-1987   **(4)** <u>Copywriter, International Food Advertisers, Inc.</u> Wrote wide variety of copy for booklets, advertisements, and brochures describing various food products. Clients included Egg Council and three manufacturers of prepared foods.

1978-1982   **(5)** <u>Administrative Assistant and Research, Food, Inc.,</u> a concern that published three monthly magazines geared to the food industry. In addition to my secretarial responsibilities, I gradually began to research articles and write brief pieces for these publications.

EDUCATION

(6) 1978, B.A. Crowell University, liberal arts
(7) 1977, Associate's degree; Shay Junior Colege, liberal arts
(8) Have taken various night-school courses:

> Copywriting, Indiana University Extension
> Business Law, Indiana University Extension
> Business Administration and Marketing, Advertiser's Institute
> Marketing Strategies, Advertiser's Institute

(9) CONTACT

Phone: 619-444-0568
2317 Vista del Robles
La Mesa, CA 94301

References and samples available upon request.

General Issues

Resume is in the wrong format for someone who wants a very specifically defined position. She needed to dig back into her job-keeping records to define several capabilities and accomplishments that would strengthen the new targeted format.

1. "Objective" should be "Job Target."
2. "Work History" is weak; "Employment History" in new positioning is stronger.
3. No city listed for employer.
4. No city listed for employer.
5. No city listed for employer
6. Her education was so long ago, graduation date should not be used.
7. Associate degree date does not correspond to graduation date. If she graduated with a B.A. in 1978, she should have received her A.A. in 1976. Either way, this kind of error could cause a potential employer to question her veracity and wonder "What else is not true?" Best solution: Given the fact she has achieved her B.A., no other school need be listed.
8. Needs stronger format to list postdegree courses.
9. Contact information is in the wrong place.

SONYA McREADY
2317 Vista del Robles
La Mesa, CA 94301
(619) 444-0568

JOB TARGET: Copywriting position in which I can use my experience in food marketing.

CAPABILITIES:
* Creative idea generator
* Excellent organization, communication, and writing skills
* Equally effective working in self-managed projects or as a team member
* Extremely dependable in completing projects on-time
* Extensive research capabilities—online and in libraries

ACHIEVEMENTS:
* Wrote for five major food accounts in current position—Idaho Potato Board, Oregon Fisheries Association, Great Western Meat Council, California Avocado Advisory Board, and Southwest Grain Board

* Researched and wrote California Avocado Advisory Council Cookbook

* Directed two junior copywriters on my team in producing two monthly client magazines, and various newsletters, print ads, and brochures.

* Won "Lulu" from Los Angeles Ad Women for Egg Council Campaign

EMPLOYMENT HISTORY

1987-Present	Senior Copywriter, Burton Advertising, San Diego, CA
1982-1987	Copywriter, International Food Advertisers, Los Angeles, CA
1978-1982	Administrative Assistant and Researcher, Food, Inc., Wayne, IN

EDUCATION

B.A. Liberal Arts, Indiana University, Indianapolis, IN

Night courses at Indiana University and Advertisers Institute, Los Angeles, in:
*Copywriting *Business Law *Business Administration and Marketing

References and samples available upon request.

PROBLEM RESUME: ADVERTISING MANAGER

Margaret Fuller
543 Eagle's Nest Drive
White Plains, NY 10602
(914) 987-6432

Objective: To provide a large manufacturing, wholesaling, or retailing corporation with creative, imaginative, sales-building advertising programs. (1)

Advertising and Promotion Experience

Advertising Manager: Contact Optical Centers, Greenwich, CT 1987-Present
This $25 million company has 29 outlets in Connecticut, New Jersey, and New York providing eye examinations, eyeglasses, and contact lenses at low cost. It has recently engaged in an aggressive expansion into three mid-Atlantic states with 15 new mall locations. My duties are: (2)

*Planning all advertising and promotion campaigns in consultation with president and sales manager.

*Conducting all newspaper, billboard, and direct mail advertising on a budget of $1.25 million.

*Managing a staff of ten graphic designers, copywriters, and assistants.

*Guiding the expansion program, including designing special promotions.

*Utilizing Macintosh Desktop Publishing for graphics in a variety of artwork genres. (3)

Since joining this company in 1987, sales have increased 200 percent. Tests show that my promotions draw sales of 5 to 8 percent. Desire change because I am ready for wider, higher-echelon responsibility. (4)

Advertising Manager: Bronners Department Stores, Columbus, OH 1984-1987
This midwestern family department store chain grosses $140 million a year. As advertising manager, I made daily, Sunday, holiday, and special event layouts for newspaper ads and supervised preparation of copy and production; I supervised the production of radio and TV (5) announcements. I prepared all envelope stuffers for continuous campaign mailings to charge customers. Left to gain higher responsibility at Contact Optical Centers. (6)

Copywriter: Advance Advertising, Battle Creek, MI, 1982-1984
This agency had billings of over $10 million, including the point-of-purchase contract for Famous Breakfast Foods, the company's largest account. I wrote all copy for the Famous displays. Left for position of greater managerial responsibility at Bronners.

Professional Training in Advertising and Sales ⑧

<u>Certificate in Direct Mail Advertising</u> Academy of Direct Mail Advertising, Advertising League of Ohio, 1985. While working for Bronners Department Stores, I took an eight-week course and seminar in direct-mail advertising. My model letters won first prize and earned me a $500 award.

<u>Certificate in Sales Promotion</u> School of Continuing Education, University of Bridgeport, Bridgeport, CT, 1988. While working for Contact Optical, I completed three continuing education courses in sales, sales promotion, and advertising, which led to a special certificate. These courses were conducted by leading authorities.

<u>Bachelor of Science, Marketing</u>, Eastern Michigan University, Ypsilanti, MI, 1982. Trained in all phases of marketing with heavy emphasis on advertising and promotion.

<u>References</u> ⑨

Mr. Robert Yotta, President; Contact Optical Centers, 1200 Corporate Towers, Greenwich, CT 06830 (203) 866-1111

Ms. Angela Bronner, Vice-President; Bronners Department Stores, 2400 Bronner Building, Columbus, OH 43216

Mr. Frederick Neuman, Managing Partner; Advance Advertising, 250 Otsego Way, Battle Creek, MI (616) 663-3636.

General Issues

Fuller presents herself as a creative communicator, capable of successfully promoting her employer's products. But her resume fails to prove it. It's not creative or concisely written. Though she emphasizes accomplishments in places, in general it's too long, too wordy, and contains unnecessary information. Its blocks of copy make it uninviting to read and the type is like a gray mass on the page. Dates are buried within paragraphs, making it difficult to tell at a glance where she worked at a certain time. A one-page resume with a chronological linear format will tell her story more concisely, leading to her next upward career move.

1. Omit objective. Use a summary of qualifications.
2. Emphasize results, not responsibilities. Emphasize her own effectiveness.
3. Don't overuse buzzwords. "A variety of artwork genres" is too abstract and Macintosh Desktop Publishing is too advertising specific.
4. Empasize individual accomplishments. Be more specific with percentages.
5. See #4.
6. Don't give reason for leaving previous job.
7. Keep resume to one concise page.
8. Education section is too long, it "I's" the reader too much and is too personal.
9. Don't include names of references.

Margaret Fuller
543 Eagle's Nest Drive
White Plains, New York 10602
(914) 987-6432

Summary

Advertising manager with eight years in-house and agency experience with increased responsibility and scope. Idea person with proven ability to manage and lead a creative staff. Successful sales builder with media, direct-mail, point-of-purchase, and special promotions record.

Experience and Accomplishments

1987-present   Contact Optical Centers; Greenwich, Connecticut
Advertising manager with in-house staff of ten and annual budget of $1.25 for 29-unit chain of optical centers grossing $25 million per year.
- Reduced media budget while significantly increasing exposure by instituting in-house agency.
- Negotiated, purchased, and installed five-station Macintosh Desktop publishing network, which resulted in time and cost savings over traditional layout and typesetting while increasing in-house graphic and artistic capabilities.
- Created and controlled $.5 million promotional campaign to herald 15-unit expansion. As a result, first-quarter sales in new stores were 10 percent ahead of projections.

1984-1987   Bronners Department Stores; Columbus, Ohio
Advertising manager for midwestern family department store chain grossing $140 million per year.
- Increased sales 8 percent in one year as a result of intensive direct-mail campaign to credit customers.

1982-1984   Advance Advertising; Battle Creek, Michigan
Copywriter for Famous Breakfast Foods point-of-purchase promotions. Created nationwide "Tommy the Tiger" contest, which was credited with increasing market penetration by 5 percent.

Education

B.A. Marketing, Eastern Michigan University
Continuing education courses in direct mail and sales promotion

Special Skills and Knowledge

Aldus PageMaker, Microsoft Word, MacDraw, and MacWrite programs. Skilled in operating others.

References Available Upon Request

Bridget K. Lyons   Bank Attorney

① 227 Upper Avenue, Des Moines, Iowa 50309 (515) 865-9158

Objective: To serve in the legal department, personal trust department, or
② corporate trust department of a large commercial bank.

Education: J.D. Drake University School of Law (evenings) 1989
Member of Bar, State of Iowa, 1990

③ B.S. Economics, Iowa State University, 1985
(Earned tuition and maintenance as night-shift typist for legal
department of First National Bank of Iowa)

Experience: Trust Officer, City Bank, Des Moines, Iowa, 1989 to present.

④ Main duties: Interview persons requesting personal trust information; design personal trusts, administer trusts, and speak to community groups

Manager, First National Bank, Des Moines, Iowa, 1985-1989

⑤ Main duties: Hired as economic analyst, 1985; promoted to
assistant manager, Corporate Trust Department, 1987, then to
contract review manager, legal manager, legal department, 1988.

⑥ Left for more responsible position in larger bank.

Clerk Typist, Legal Department, First National Bank, 1981-1985

⑦ Reason for Leaving: graduated from college, took full-time
position

References: Furnished upon request

RESUME KEY FOR BRIDGET K. LYONS: ATTORNEY

General Issues

Lyons won't budge a judge at the human resources hiring court. She failed to elaborate on her achievements of five years in banking. The sketchy information doesn't give any idea about her leadership skills, managerial ability, or even professional credentials. She also failed to highlight dates and job titles to make it easy to follow her career. She gives an objective, reasons for leaving, and lists education first, rather than last. As a lawyer, she should know better!

1. Reformat identification block.
2. Omit objective; change to summary.
3. Position education at bottom.
4. Too weak; needs stronger results in her position.
5. See #4.
6. Omit reason for leaving.
7. See #6.

Bridget K. Lyons
227 Upper Avenue
Des Moines, Iowa 50309
(515) 865-9158

Career Summary: Attorney with economics background (including analysis), corporate trust management, and personal trust management achievement in major banks.

Professional Achievements

1989-present   **Trust Officer, City Bank, Des Moines, Iowa**

Manager of personal trust staff of 15, monitoring investments of $700 million. Established department policies, developed marketing and customer service programs, instituted community relations program, administered trusts, and controlled income to the bank in excess of $50 million. My policies and efforts resulted in increased visibility for new department and new revenue to the bank of $20 million in the first year of the department's operation.

1985-1989   **First National Bank, Des Moines, Iowa**
Economic Analyst, 1985-1987

- Identified economic trends and environmental conditions affecting the bank and prevented losses from overemphasis in agricultural loans. My proposals and recommendations resulted in the diversification of the bank portfolio and overall strengthening of its financial picture. As a result of my success, promoted to

Assistant Manager, Corporate Trust Department, 1987-1988

- Revised policies and established a trust account review system that resulted in an average 2 percent per year increase in return to clients and 24 percent annual increase in department revenue. Upon graduation from law school, promoted to

Contract Review Manager, Legal Department, 1988
- Created new, more effective policies for contract review, rewrote policy manual for contract preparation, and hired members into new department of contract writers.

Education

J.D. Drake, University School of Law (evenings)

Member of the State Bar of Iowa

B.S., Economics, Iowa State University

References

Furnished upon request.

Gail Anderson
542 Pine Street
San Francisco, CA 94576
(415) 685-6784

① OBJECTIVE   Financial management position with personal and professional growth potential

**PROFESSIONAL
EXPERIENCE**   Coopers & Lybrand

②   I have been employed by Coopers & Lybrand for over five years. As an audit supervisor, I am responsible for all facets of audit and non-audit engagements including planning, budgeting, report preparation, and supervision of senior and staff accountants. I have client service responsibilities for public and privately held companies in the retail food service, heavy construction, and manufacturing industries.

Specific experience and achievements with Coopers & Lybrand include:

*Involvement in the initial public stock offerings for five clients, which included discussions with and reporting to the Securities and Exchange Commission.

*Supervision of services for a developmental stage steel mill, which entailed becoming familiar with complex financing arrangements.

*Experience in the use of microcomputers and the input and output data from all sizes of EDP shops.

*Performance of special litigation support procedures.

*Consultation in the sale of client businesses, which included contacting potential buyers.

**PROFESSIONAL
ORGANIZATIONS**   American Institute of Certified Public Accountants
California Association of Certified Public Accountants

EDUCATION   Bachelor of Science in Business Administration
③   Major in Accounting
University of San Francisco, 1987

PERSONAL DATA   Single
④   Excellent health
Personal interests include sailing, biking, and tennis

REFERENCES   Furnished upon request.

General Issues

Graphically, the look of Anderson's resume is too long and linear. Type needs to be bigger. With only one employer, a chronological resume is out of the question. The functional resume, with its breakout of specific achievements, is best at this point in her career.

1. "Objective" should be omitted. It says nothing specific.
2. Resume format is incorrect.
3. In this case, though she has been out of school more than five years, the fact of her relatively recent graduation explains why, thus far, she has only had a single job.
4. Omit personal data.

GAIL ANDERSON
542 Pine Street
San Francisco, CA 94576
(415) 685-6784

EXPERIENCE

Audit Supervision: Direct all facets of audit and nonaudit engagements including planning, budgeting, report preparation, and supervision of senior and staff accountants. Recently supervised services for developmental stage steel mill involving complex financing transactions.

Client Servicing: Oversee financial reports for diversified clientele including public and privately held companies in retail food service, heavy construction, and manufacturing. Recently consulted with food service client for sale of business including potential buyer contact.

Stock Offerings: Assisted in development of public stock offerings for five clients. Coordinated ongoing discussions with Securities and Exchange Commission.

Litigation Support: Interfaced with client attorneys in seven-figure civil damages suit.

Data Management: Developed criteria for external EDP data support shops. Personally familiar with IBM PC using Excel and Lotus 1-2-3.

Professional Skills: Highly attentive to detail. Strength in evaluating potential growth strategies. Able to plan, direct, and coordinate complex financial programs.

EMPLOYMENT HISTORY

1987 - Present Coopers & Lybrand, San Francisco, CA

PROFESSIONAL ORGANIZATIONS

American Institute of Certified Public Accountants
California Association of Certified Public Accountants

EDUCATION

BS in Business Administration, Accounting Major
University of San Francisco, 1987

REFERENCES

Complete references furnished upon request, once mutual interest has been established.

(1)

Herman F. Fox

346 N. Framingham Lane
Portland, OR 98431   (2)
(516) 883-4531   OBJECTIVE

                                TO SERVE MAJOR INDUSTRIAL CORPORATION
                                AS FINANCIAL-TECHNICAL AUDIT MANAGER
                    (3)
            EXPERIENCE AS OPERATIONAL AUDITOR
                                (4)
Fifteen years of heavy operational auditng experience with major
(5) chemical products manufacturer and large power and gas public
utility, after thorough training in engineering(BSEE) and accounting
(MBA).
                (6)   (7)
AUDIT   Nine years' experience with Brown Chemical, Inc. as   (8)
MANAGER   Operational Auditor (1982-1988)and Audit Manager (1988-
            present). I designed the basic systems and procedures for
enlarging the financial audit of this company to a full audit that
included engineering design, production, and control by
computerization using both the IBM PC and Lotus 1-2-3.

My system integrated perfectly with the financial audit and
qualified me for the top audit post when incumbent Audit Manager
retired. Responsible for both the financial audit and production
audit of the company and reported directly to the President.
                    (9)   (10)
INTERNAL   Six years' experience with Metropolitan Utility as
AUDITOR   Internal Auditor (1976-1982). I reviewed all accounts,
            contracts, disbursements, and field performance of
contract analysis (vegetation, gas, and wiring). My unique
combination of engineering and accounting made my writing procedures
for this type of audit, and the resultant report, possible. These
procedures are still used by this utility.

                        EDUCATION

Master of Business Administration (MBA), Leonard N. Stern School of
  Business, Graduate Division, New York University, 1976.(11)

Bachelor of Science, Electrical Engineering (BSEE), Albany Technical
  Institute, Albany, New York, 1970.(12)

                        MILITARY

First Lieutenant, United States Army, Signal Corps, Instructor in
  Electronics, Fort Monmouth. Three years of service.

                        REFERENCES

Complete references will be supplied on request.

RESUME KEY FOR HERMAN F. FOX: AUDITOR/TECHNICAL

General Issues

Though Fox may be well qualified with his combination of financial and technical expertise, you'd never know it by this confusing resume. Your eyes wander all over the page trying to find some continuity. Aside from poor layout, the lack of identifying major paragraphs and the mixed-up nature of his explanatory information would lead any human resources specialist to say "pass" after one glance.

1. Incorrect positioning of identifying information.
2. "Objective" needs to be folded into strong summary for this unique candidate.
3. Wrong title.
4. Misspelled word.
5. This paragraph needs to be part of the summary.
6. No apostrophe needed.
7. Incorrect positioning of dates.
8. See number 7.
9. No apostrophe needed.
10. See number 7.
11. Shorten degree. Also no dates needed for graduation; it was too long ago.
12. See number 11.

HERMAN F. FOX
346 N. Framingham Lane
Portland, OR 98431
(516) 883-4531

CAREER SUMMARY: Fifteen years of heavy operational auditing experience with major chemical products manufacturer and large power and gas public utility. Combination of backgrounds in engineering and accounting prepare me for position as financial-technical audit manager with major industrial corporation.

PROFESSIONAL EXPERIENCE

1992 - Present   **Brown Chemical, Inc., Portland, OR**

Audit Manager (1988 - Present)

Operational Auditor (1982 - 1988)

Report to President of this diversified chemical products firm. Responsible for day-by-day and extended financial and production audits. Designed all basic systems and procedures to enlarge scope of financial audit to include engineering design, production, and control. Created computerized system using both the IBM PC and Lotus 1-2-3.

Accomplishments: 20% greater efficiency in production scheduling.
Timely financial analysis led to 15% savings in overtime over 3-year period.

1976 - 1982 **Portland Metropolitan Gas & Electric, Portland, OR**

Created and wrote procedures manual for combination audit still in use today. Drafted format for audit that encompasses all accounts, contracts, disbursements, and field performance of contract analysis (vegetation, gas, and wiring). Oversaw all aspects of financial analysis.

EDUCATION

MBA, Leonard N. Stern School of Business, Graduate Division, NYU
BSEE, Albany Technical Institute, Albany, NY

MILITARY

First Lieutenant, United States Army, Signal Corps, Electronics Instructor, Fort Monmouth, NJ

COMPLETE REFERENCES WILL BE SUPPLIED ON REQUEST

①
Henry Watson
8343 West Grover Street
St. Louis, MO 38192
(314) 858-0411

②

EXPERIENCE:

③ 8/88
to Present

Assistant Vice President/Credit Officer
Commerce Bank Missouri, St. Louis, MO

Obtain credit and financial information necessary to assess
the financial risk of commercial and installment loans.
Analyze the information, draft loan reports, and make
appropriate recommendations. Interview loan applicants to
determine credit qualifications and borrowing needs.
Provide deposit and loan services to existing customers.
④ Prepare monthly loan reports.

4/86
to 7/88

Assistant Vice President/Loan Administration
St. Louis National Bank, St. Louis, MO

Directed the preparation of commercial, consumer, real
estate, and construction loan documentation. Assisted the
senior credit officer in supervising loan compliance and
regulatory issues applicable to bank lending. Wrote
material governing lending practices, policies, and
procedures. Exercised approval authority within established
limits of $50,000 to individuals and $100,000 to
businesses. Provided collection assistance for commercial,
⑤ consumer, and real estate problem loans.

11/84
to 4/86

Assistant Bank Examiner
Regional Bank Systems, St. Louis, MO

Assisted in bank examinations and loan audits. Reviewed
and analyzed problem loans.

5/83
to 10/84

Loan Officer
Commerce Bank Missouri, St. Louis, MO
Interviewed loan applicants, analyzed their requests, and
made appropriate recommendations.

EDUCATION: B.A. Business Administration, Washington University ⑥

COMMUNITY: Insurance Committee - St. Louis Chamber of Commerce
Board of Directors - Bankers Club of Missouri

RESUME KEY FOR HENRY WATSON: BANK ASSISTANT VICE PRESIDENT/CREDIT OFFICER

General Issues

There isn't a whole lot wrong with this resume, outside of the fact that it's dull. The verbiage could use brightening, and accomplishments are needed. Right now, Watson looks like a run-of-the-mill candidate. Better type layout would help, too.

1. Capitalize name.
2. Lacks strong selling summary.
3. Omit months.
4. Achievements are missing.
5. Achievements are missing.
6. Lacks location of university.

HENRY WATSON
8343 West Grover Street
St. Louis, MO 38192
(314) 858-0411

Problem solving banking professional with over 11 years' experience. Using interpersonal skills, able to sort "wheat from chaff" to determine qualified loan applicants. "People" orientation combined with financial analytical skills.

EXPERIENCE:

1988
to present

Assistant Vice President/Credit Officer
COMMERCE BANK MISSOURI, St. Louis, MO
Responsibilities: Obtain credit and financial information from consumer and business loan applicants. Analyze data, draft reports, and make appropriate recommendations. Build customer good-will through one-on-one deposit/loan services. Produce monthly problem reports.
Achievements: Built equity home loan business by 20% in 1993. Sucessfully maintained past due accounts under 3%.

1986-1988

Assistant Vice President/Loan Administration
ST. LOUIS NATIONAL BANK, St. Louis, MO
Responsibilities: Directed preparation of commercial, consumer, real estate, and construction loan documentation. Assisted in supervision of loan compliance and regulatory issues. Approved loans with established limits of $50,000 (individual) and $100,000 (business). Oversaw collection for all problem loans. Wrote standards and practices for lending policies
Achievements: Expanded business banking segment by 35% during my tenure. Attained up to 60% collection ratio.

1984-1986

Assistant Bank Examiner
REGIONAL BANK SYSTEMS, St. Louis, MO
Responsibilities: Assisted in bank examinations and loan audits. Reviewed, analyzed, and reported on problem loans.

1983-1984

Loan Officer
COMMERCE BANK MISSOURI, St. Louis, MO
Responsibilities: Interviewed loan applicants; made recommendations.

ORGANIZATIONS:

Insurance Committee—St. Louis Chamber of Commerce
Board of Directors—Bankers Club of Missouri

EDUCATION:

B.A. Business Administration, Washington University, St. Louis, MO

Anne B. Morrissey
28346 Angeles Vista Road
Westlake Village, CA 91302
(805) 734-6893

(*1*) <u>Objective</u>   A position of responsibility in a team-oriented atmosphere offering long-term advancement.

(*2*) <u>Education</u>   (*3*) 1984 B.A. California State University, Northridge
1982 A.A. Los Angeles Valley College

<u>Job History</u>
1991 to   <u>Queen and Alton</u>, CPAs, Calabasas, CA
present   Position: Staff Accountant
               Responsible for payroll, accounts payable, and accounts receivable
               for 25 restaurants with 50 to 80 employees per store. **Proficient** (*4*)
               in AccPac software. Supervised and trained new hires. Other skills
               include 10 key by touch and 50 WPM typing speed.

1988 to   <u>Management Team Services, Inc.</u>, Westwood, CA
1991   Position: Staff Accountant
               Part of a team of six staff members managing corporate clients.
               Controlled accounts for 12 clients. Reconciled monthly bank
               statements and did month end closing reports for corporate tax
               consultants. Full-charge bookkeeping duties included accounts
               payable, accounts receivable, automated payroll, and health insur-
               ance plans.

1985 to   <u>Charles Filibrow, Inc.</u>, Encino, CA
1988   (*5*) Position: Accounting Clerk
               Assisted full-charge bookkeeper. Heavy general ledger activity for
               accounts receivable, payroll, and month end reports.

1983 to   <u>FXR Entertainment Group, Inc.</u>, Woodland Hills, CA
1985   Position: Staff Accountant
               Member of an 8-person team, which managed 150 limited enter-
               tainment partnerships. Responsible for accounts payable, accounts
               receivable, payroll, general ledger, month end reports.

1980 to   (*6*) <u>Hennessey Motors</u>, Van Nuys, CA
1983   Position: Accounting Clerk
               Responsible for accounts payable, accounts receivable, payroll,
               heavy data entry, and month end general ledger reports.

RESUME KEY FOR ANNE B. MORRISSEY: BOOKKEEPER

General Issues

Resume is neat and tidy reflecting an orderly bookkeeper's mind. Unfortunately, it has faulty parallelism, lacks strong verbs, and is in incorrect order.

1. Objective should be replaced by strong summary.
2. Education is in wrong place.
3. Graduation dates need to be removed.
4. Skills need to be in separate section.
5. Since she assisted full-charge bookkeeper, she can lay claim to the stronger job title of assistant bookkeeper.
6. Earliest position unnecessary.

Anne B. Morrissey
28346 Angeles Vista Road
Westlake Village, CA 91302
(805) 734-6893

Summary: Over 10 years progressively more responsible positions in all phases of bookkeeping and office management. Determined self-starter who maintains all deadlines and keeps deadly accurate records. Not intimidated by heavy work volume.

Employment:

1991 to present

Queen and Alton CPAs, Calabasas, CA
Staff Accountant
Direct all financial management for 26 restaurants with 50 to 80 employees per store. Responsible for payroll, accounts payable, and accounts receivable. Supervise and train new hires.

1988 to 1991

Management Team Services, Inc., Westwood, CA
Staff Accountant
Controlled accounts for 12 clients as part of a team of six. Reconciled monthly bank statements. Completed month end closing reports for corporate tax consultants. Performed full-charge bookkeeping duties: accounts payable, accounts receivable, automated payroll, health insurance plans.

1985 to 1988

Charles Filibrow, Inc., Encino, CA
Assistant Bookkeeper
Coordinated heavy general ledger activity for accounts receivable, payroll, and month end reports.

1983 to 1985

FXR Entertainment Group, Inc., Woodland Hills, CA
Staff Accountant
Managed 150 limited entertainment partnerships as part of eight-person team, with responsibilities that included accounts payable, accounts receivable, payroll, general ledger, and month end reports.

Skills: AccPac software; 10 key by touch; 50 wpm typing speed.

Education: B.A., Business Administration
California State University, Northridge

References: Available upon request.

Consuelo V. Alvarez
5465 Sandia Way, Albuquerque, NM 87114
(505) 345-6723
HOSPITAL ADMINISTRATOR

① Professional Objective

To serve as Chief Administrator of a large general hospital.

Experience

SANDIA GENERAL HOSPITAL   ALBUQUERQUE, NM
Acting Chief Administrator, 1989 to present

Direct all activities of this 200-bed hospital, reporting to the
Board of Governors, in the absence of permanent Chief
Administrator, who will return to hospital after two years'
sabbatical in December.

Oversee hiring, training, and management of all nonprofessional
personnel. Directed hospital budget, instituted cost-containment
measures, expedited collections and third-party reimbursement.
Initiated private capital campaign, which has resulted in $2
million increase in hospital endowment. Represent hospital in all
community affairs. Established new community health projects.

ALBUQUERQUE VETERANS HOSPITAL   ALBUQUERQUE, NM
Assistant Hospital Administrator, 1985-89
Designed, implemented, and administered new computerized admissions
system when hospital was renovated in 1985. Put in charge of $20
million three-year renovation project. Lobbied Congress and the
Veteran's Administration for funding, solicited bids for project,
negotiated contract, and worked closely with contractors to bring
renovation project in at bid and on schedule. As hospital's
training officer, interpreted Veteran's Administration training
policies and developed professional and nonprofessional improvement
programs.

Education

UNIVERSITY OF TEXAS   AUSTIN, TX

Master of Hospital Administration, 1982-85

② Three-year program, including one-year administrative residence at
M.D. Anderson Cancer Center, Houston, TX. Graduated with honors.
Academic program included the following courses:

Hospital Management
Medical Care Administration
Computerized Hospital Accounting Systems
Case Studies in Hospital Design
Community Relations and Fund-Raising
Hospital Budgeting
Professional and Nonprofessional Personnel Management

UNIVERSITY OF NEW MEXICO   ALBUQUERQUE, NM

<u>Bachelor of Science,</u> (Biology Major)   1978-82

Dean's List every semester. Graduated <u>cum laude</u> with 3.5 average. Attended on National Merit full-tuition scholarship. Worked at Community Hospital as admissions clerk part-time to pay living expenses, as a result of which work I decided on graduate training and a professional career in hospital administration.

SPECIAL SKILLS & KNOWLEDGE: Speak and read fluent Spanish & English

Persuasive public speaker and presenter

RESUME KEY FOR CONSUELO V. ALVAREZ: CHIEF HOSPITAL ADMINISTRATOR

General Issues

Alvarez has it almost right. She shows that she has excellent professional and academic credentials. But what she needs is a resume that reflects a knowledge of the market and what the consumer will buy. The revision shows the writer knows that the hospital's board of directors is looking for an administrator who can do more than routine management tasks—one who can successfully motivate the hospital's professional and administrative staff to accomplish their goals.

1. Omit objective.
2. Shorten and tighten education.

RESUME SOLUTION: CHIEF HOSPITAL ADMINISTRATOR

Consuelo V. Alvarez
5465 Sandia Way
Albuquerque, New Mexico 87114
(505) 345-6723

Summary: Graduate degree and five years' experience in hospital administration, including acting chief administrator. Effective manager with unique ability to train, motivate, and direct people. Successful record in capital improvement, management of renovation projects, installation of computerized management systems, and development of training problems. Problem solver with human resources skills.

Accomplishments

1989-present   *Acting Chief Administrator*
                                            Sandia General Hospital, Albuquerque

Directed all activities of this 200-bed hospital, reporting to the Board of Governors in the absence of permanent Chief Administrator during his two-year sabbatical.

- Managed all nonprofessional personnel. Developed and administered hospital budget, instituted cost-containment measures, expedited collections and third-party reimbursement. Initiated private capital campaign, which has resulted in $2 million increase in hospital endowment. Represented hospital in all community affairs. Established three new community health projects, which significantly enhanced hospital's presence in community.

1985-1989   *Assistant Hospital Administrator*
                                            Albuquerque Veterans Hospital

- Designed, implemented, and administered computerized admissions system.
- Directed $20-million three-year renovation project. Lobbied Congress and the Veteran's Administration for funding, solicited bids for project, negotiated contract, and worked closely with contractors to complete renovation project at bid, on schedule.
- As hospital's training officer, interpreted Veteran's Administration training policies. Developed effective professional and nonprofessional improvement programs.

Education: *Master of Hospital Administration,* University of
                                           Texas
                                           (Degree of Distinction)
                                           Bachelor of Science, University of New Mexico
                                           Biology Major, graduated *cum laude*

Special Skills and Knowledge: Speak and read fluent Spanish and English
                                           Persuasive public speaker and presenter

MICHELLE BORDERS
(1) 573-A Warren Avenue
Spokane, WA 95712
(206) 812-9753

(2) JOB INTERESTS: Office Manager, Payroll, Supervisor, Communications Clerk, Accounting Clerk, Billing Clerk.

CURRENT EMPLOYMENT: (3)
BLAIR UNITED SCHOOL DISTRICT, # 129 (4)
Dates of Employment: October 1992 to Present (5)
Position: Payroll Clerk

Job Function:
To process the monthly payroll and all related procedures. I process payroll for approximately 1300 employees, nonunion and administrative employees. I enter all information relating to the employee's pay, benefits, state retirement, union
(6) affiliation, account distribution, TSA's, direct deposit, etc. In addition I balance and reconcile all annuities, insurance's and the like. I prepare direct deposit and the tax deposits. I process, answer, and carry out garnishments, IRS levies, child support orders, etc. I work with the ESD "Versaterm" computer system and Word, Works, and Excel on the Macintosh system. My annual salary is $23,127.00. I prefer you do not contact my current employer. (8)   (7)

KAISER PERMANENTE OF THE NORTHWEST (9)
Dates of Employment: August 1990 to September 1992 (10)
Position: Payroll Specialist

Job Function:
To pay employees in an accurate and timely manner, according to federal and state laws, complex union contracts, and company policies. We had over 6000 employees. I handled the payroll for approximately 1000 employees personally. I worked with two entity company (Kaiser Permanente Health Plan and Kaiser Permanente Hospitals). I worked with multiple unions, nonunion, salaried, exempt, nonexempt hourly, and administrative personnel. I processed COMPLEX personnel action forms relating to a person's pay, benefits, union contract, company policy,
(11) etc. I processed time sheets and audited for compliance with wage and hour laws, contracts, company policies, etc. In addition, I processed manual checks for adjustments, awards, bonuses, retro payments, grievances, legal settlements, etc. I set up union dues and processed workmen's compensation claims. I balanced and reconciled the bi-weekly report for the hospitals, the garnishment and child support reports, and the TSA report on a bi-weekly basis. In addition, I answered all relating questions and worked with employees, supervisors, compensations, human
(12) resources, outside agencies, etc., to gather and relate information. My rate of pay was $11.49 per hour and my supervisor was Harriet Forman. Harriet may be reached at 206-358-0063.   (13)

PACIFIC POWER AND LIGHT (14)
Dates of Employment: July 1987 through August 1990 (15)
Job Position: Payroll Specialist/Accounting Clerk III

Job Function:
To process the payroll for approximately 8000 employees in seven states, multi unions and nonunion personnel. I audited and corrected time tickets, processed (16) COMPLEX batch on both the MSA Payroll system and an in-house system. I manually figured and wrote or authorized checks for new hires, terminations, etc. I dealt with tax issues in seven states as well as six different unions, nonunion, and administrative issues. I balanced and reconciled the salary advance account and employee purchase accounts. In addition, I processed and balanced the deferred compensation account and all employee bonuses. I handled all distribution of paychecks and related issues. My salary was $1608.00 per month. My supervisor was Stanley Billingham. He has since retired. (17)

(18)
(19) CURRENT EDUCATION:
March 1990 Associate in Applied Science in Business
Administration Degree
Lewis College, Vancouver, Washington

(20) June 1983 Diploma. Highlands High School, Vancouver, Washington

REFERENCES
Available upon request.

General Issues

With the voluminous detail Borders gives about each position, you can see she's extremely detail oriented. That's a plus in her career path; it's just that employers don't need to see such minutiae. The large copy blocks are a turnoff and very hard to read. Cutting through the dead wood and omitting such gaffes as salary and supervisor will go a long way toward helping Michelle get on with her career.

1. Center identifying information.
2. Omit job interests and add "Summary."
3. Omit "Current."
4. Need location of employer.
5. Omit month.
6. Edit, edit, edit.
7. Omit salary.
8. Omit supervisor's name.
9. See #4.
10. See #5.
11. See #6.
12. See #7.
13. See #8.
14. See #4.
15. See #5.
16. See #6.
17. See #7.
18. See #8.
19. See #3.
20. Omit high school credentials; college degree supersedes.

MICHELLE BORDERS
573-A Warren Avenue
Spokane, WA 95712
(206) 812-9753

SUMMARY: Multifaceted, detail-oriented payroll clerk with seven years' experience. Handled all responsibilities of payment for thousands of employees at each position. Trained in complex issues dealing with unions and workers' compensation. Ready to move into management.

EMPLOYMENT

1992 - Present   BLAIR UNITED SCHOOL DISTRICT, #129, Blair, WA
Payroll Clerk
Process monthly payroll for approximately 1300 employees, union, nonunion, and administrative. Perform all related procedures: enter all information related to pay, benefits, state retirement, union affiliation, account distribution, TSA's, direct deposit. Process direct deposit, reconcile all annuities and insurance. Coordinate government related functions: garnishments, IRS levies, child support orders, etc.
Computer Systems: ESD "Versaterm" and Word, Works, and Excel on Macintosh.

1990 - 1992   KAISER PERMANENTE OF THE NORTHWEST, Spokane, WA
Payroll Specialist
Personally managed payroll processing for 1000 of Kaiser's 6000 employees—union, nonunion, salaried, exempt, nonexempt hourly, and administrative personnel. Handled complex forms relating to pay, benefits, union contract issues, wage and hour laws, and company policies. Set up union dues and processed workers' compensation claims. Balanced and reconciled the following reports: hospital weekly report, garnishment and child support, and TSA. Processed manual checks for adjustments, awards, bonuses, etc. Researched information and answered related questions from employees, management, human resources, outside agencies.

1987 - 1990   PACIFIC POWER AND LIGHT, Spokane, WA
Payroll Specialist/Accounting Clerk III

Processed payroll for approximately 8000 employees in seven states, multiunion and nonunion personnel. Audited and corrected time tickets, processed complex batch on both MSA Payroll system and in-house system. Dealt with tax issues in seven states, plus six unions and numerous administrative issues. Processed and balanced deferred compensating account and all employee bonuses. Physically handled distribution of paychecks.

EDUCATION
1990   Applied Science in Business Administration Degree, Lewis College, Vancouver, WA

REFERENCES
Current employer does not know of job search. Other references available upon request.

Roger Silvers
8311 West 1st Street
Oxnard, CA 97321
(805) 982-5312

① JOB OBJECTIVE
To obtain a responsible position in Computer Operations with a defined career and continued professional growth opportunities.

EXPERIENCE
③
② February 1992 - Present, LAIDED OFF

October 1988 - February 1992, DP Tape Librarian
HUGHES AIRCRAFT CO., Los Angeles, Ca.
As Librarian I'm responsible for the operation of the Data Processing Library System. Monitoring, updating, and correcting **④** TMS records. Staging microfiche processing and system backups for offsite storage. I write clist to enhance my job performance as librarian and I use the Document Composition Facility (SCRIPT).

August 1986 - October 1988, Computer Operator
⑤ HUGHES AIRCRAFT CO., Los Angeles, Ca. **⑥**
Was operating a DEC VAX 11/780 and VAX 8600 computer and peripheral devices, using UNIX and VMS operating system. Did general system maintenance, incremental dumps, updating libraries for Ingres Data Base, used Q-Calc spreadsheets for reports and MS Macros for creating forms.

April 1984 - August 1986, Data Entry Operator
⑦ HUGHES AIRCRAFT CO., Los Angeles, Ca.
Inputing and verifying data.

⑧ October 1979 - April 1984, Data Entry Operator
PDQ DATA: Los Angeles, Ca.

⑨ May 1979 - October 1979, Data Entry Operator
LOUISIANA NATIONAL BANK: Baton Rouge, La.

⑩ January 1978 - May 1979, Data Entry Operator
X-Ray ASSOCIATES: Lansing, Mi.

⑪ September 1975 - January, 1978, Data Entry Operator
STATE OF MICHIGAN: Lansing, Mi.

⑫ April 1973 - September 1975, Data Entry Operator
DIAMOND REO: Lansing, Mi.

EDUCATION
Lansing Community College: Certificate in Data Entry

REFERENCES
Personal and Professional References are available on request.

RESUME KEY FOR ROGER SILVERS: COMPUTER DATA-PROCESSING SPECIALIST

General Issues

To begin a resume with an obvious layoff (especially with a typo!) is a faux pas. It would be far better to let the fact speak for itself on the resume, and address its legitimate cause—a massive slowdown in aerospace and technical fields in the 1990s in Southern California. (This is a legitimate reason for layoff.) With Silvers' extensive computer expertise, what's needed is a separate category for technical skills. Resume itself is not clear; a better format would be the linear model with focus emphasizing positions of increasing responsibility at Hughes. The way the resume is now written, the transitions are not easily made. Finally, he has too many positions listed that go back too far.

1. Omit. Replace with impressive job summary.
2. Omit any reference to layoff.
3. Misspelling.
4. Too much technolanguage in this paragraph.
5. This paragraph needs a better relationship to the above, to show continuity of employment at Hughes.
6. Equipment needs separate listing under "Technical Skills."
7. See #5.
8. Needs to be combined into one cohesive general listing as "data entry operator in a variety of positions in California, Louisiana, and Michigan."
9. See #8.
10. See #8.
11. See #8.
12. See #8.

Roger Silvers
8311 West 1st Street
Oxnard, CA 97321
(805) 982-5312

HIGHLIGHTS OF QUALIFICATIONS
* Nearly 20 years' experience of increasing responsibilities in data processing
* Expert in management of data-processing library systems
* Quickly master new software and apply full range of its capabilities
* Able to pinpoint problems and initiate realistic, workable solutions

EXPERIENCE

1984
to
1992

HUGHES AIRCRAFT COMPANY, Los Angeles, CA

Data Processing Tape Librarian (1988 - 1992)
* Managed operations of Data Processing Library System
* Monitored, updated, and corrected TMS records
* Staged microfiche processing and system backups for offsite storage

Computer Operator (1986-1988)
* Updated libraries for Ingres Data Base on a daily basis
* Created system-wide reports using Q-Calc spreadsheets
* Designed forms with MS Macros

Data Entry Operator (1984-1986)
* Input and verified data

1973
to
1984

Data Entry Operator
* Input and verified data
* Worked for data-reporting service, bank, and government institutions in California, Louisiana, and Michigan

TECHNICAL KNOWLEDGE AND SKILLS
Hardware: DEC VAX 11/780, VAX 8600, related peripherals
Software: UNIX, VMS; also special expertise in Document Composition Facility (SCRIPT), Q-Calc Spreadsheets, and MS Macros

EDUCATION
Certificate in Data Entry, Lansing Community College, Lansing, MI

REFERENCES
Excellent personal and professional references available upon request.

①
Frances Sherman
352 Trelawny Lane
Camden, NJ 07436
(609) 747-4453

OBJECTIVE

② Computer sales with a firm offering opportunities for ultimate management responsibility.

③ SALES EXPERIENCE

*Sales of CAD/CAM Software to companies in Pennsylvania, New Jersey, and New York.
*Attained 125% of my quota in orders received in assigned territory.
*Turnkey minicomputer sales in retail, wholesale, manufacturing, and service industries.
*Opened new accounts which brought in over $90,000 over a period of one year.
*Sold timesharing and remote batch services to companies of all sizes in the area of Delaware County, PA.
*Developed new business and increased sales to established accounts.

④ TECHNICAL EXPERIENCE

*Knowledge of COBOL, BASIC, and FORTRAN
*Excellent knowledge of NUMERICAL CONTROL
*Familiar with flow diagrams for chemical plant processes.
*Proficient in computer applications for accounts receivables, accounts payables, and inventory systems.

EMPLOYMENT

1992 - Present	SFN Industries
	⑤ 348 Crenshaw Parkway
	Princeton, NJ
	⑥ Position: Computer Sales
1990 - 1992	McKenna Forrester Inc.
	321 Silver Springs Road
	⑦ Erie, PA
	⑧ Position: Assistant to Sales Manager
1987 - 1990	Excalibur Construction Corp.
	⑨ 6247 W. 9th Street
	Pittsburgh, PA
	⑩ Position: Administrative Assistant

EDUCATION

1983-1987	Pennsylvania State University, State College, PA
	BA English

PERSONAL

⑪ Available for travel and/or relocation.

RESUME KEY FOR FRANCES SHERMAN: COMPUTER SALES REPRESENTATIVE

General Issues

This resume is not bad, but it could be better. With a change of phraseology (i.e., turning "Objective" into "Target") and a reformatting into a targeted resume, the results will be better focused. Note: At end, personal is quite acceptable, since it offers important information for a prospective employer without intruding on the rights of the jobseeker.

1. Reposition information block.
2. Becomes "Job Target."
3. This category is moved into "Achievements" and bullets are combined.
4. This category becomes "Technical Capabilities" and a new category is added for "Professional Skills."
5. Omit address.
6. Reposition job title.
7. See #5.
8. See #6.
9. See #5.
10. See #6.
11. This personal information is acceptable.

FRANCES SHERMAN
352 Trelawny Lane
Camden, NJ 07436
(609) 747-4453

JOB TARGET: Computer sales at manufacturer's or distributor's level, leading to ultimate management responsibility.

TECHNICAL CAPABILITIES:

* Knowledge of COBOL, BASIC, and FORTRAN
* Advanced knowledge of NUMERICAL CONTROL
* Familiar with flow diagrams for chemical plant processes
* Proficient in financial computer applications: accounts receivable, payable, and inventory systems

PROFESSIONAL SKILLS:

* Unflappable under heavy pressure
* Proven ability to exceed quotas
* Strong closer

ACHIEVEMENTS:

* Attained 125% of quota, in CAD/CAM sales throughout Pennsylvania, New Jersey, and New York.
* Opened new accounts netting over $90,000 within one year, in turnkey minicomputer sales in retail, wholesale, manufacturing, and service industries.

EMPLOYMENT HISTORY:

1992 - Present	Computer Sales Representative, SFN Industries, Princeton, NJ
1990 - 1992	Asst. to Sales Manager, McKenna Forrester, Inc., Erie, PA
1987 - 1990	Admin. Asst., Excalibur Construction Corp., Pittsburgh, PA

EDUCATION:

BA English, Pennsylvania State University, State College, PA

PERSONAL:

Available for travel and/or relocation.

REFERENCES:

Available upon request.

Jonathan P. Dean

<u>Home Address</u>
(1) 16698 Hilltop Avenue
Allentown, PA 18101
(215) 559-0090

<u>Office Address</u>
(2) Keystone Construction Corp.
Downtown Office Towers
Allentown, PA 18102
(215) 565-0331

(3) OBJECTIVE: Supervisory position in residential or commercial construction.

EXPERIENCE:

1989-present Keystone Construction Corporation, Allentown, PA
Construction Supervisor for Penns Woods residential development of
150 new homes and condominium units. Hired and supervised crew of
22, interpreted and implemented architects' building and site plans.

1987-89   Lehigh Valley Development, Mount Bethel, PA
Master Carpenter and Crew Leader. Hired as journeyman carpenter,
then promoted to master carpenter and crew leader on 500,000-
square-foot retail/office center in Mount Bethel. Directed layout,
exterior fabrication, and interior installation.

1985-87   Craft Building and Supply, Chester, PA
Journeyman carpenter assisting master carpenters on new construction
and renovation to both commercial and residential structures. Also
learned purchasing and estimating.

1981-85   Sergeant, U.S. Army Corps. of Engineers
Specialized training in construction led to equivalent of journeyman
carpenter certification. Crew leader for various construction
projects on army bases throughout United States.

EDUCATION: U.S. Army Engineering School, Maryland,
Certificate in Carpentry

Great Lakes Institute, Erie, Pennsylvania
Construction Trades Certificate (one-year)

(4) Lehigh Valley Technical High School
Four years of study in carpentry, drafting, and
mechanical design

(5) SKILLS: Fully qualified in all phases of general carpentry.
Thoroughly acquainted with lumber. Able to estimate construction
costs for developing tracts. Adept in building layout and all
surveying tools. Knowledgeable about masonry, plumbing, electricity,
heating, air conditioning, and landscaping.

RESUME KEY FOR JONATHAN P. DEAN:
CONSTRUCTION SUPERVISOR

General Issues

Dean needs to lavish a little more care on his resume in the same way he builds houses. The architecture is sound, but he should pay more attention to placement and positioning of key elements. Most importantly, instead of leaving his skills for last, he should move them up to the top in a summary that immediately highlights his abilities.

1. Place address with name in proper identification block sequence.
2. Omit office address; it's always best to talk new job business at home.
3. Omit "Objective"; include summary, instead.
4. In this case, it is beneficial to use Dean's high school background since it shows a consistent pattern of productive learning in his chosen field.
5. Use "Skills" as basis for summary near the top of the resume.

Jonathan P. Dean
16687 Hilltop Avenue
Allentown, Pennsylvania 18101
(215) 559-0090

Summary: Construction Supervisor with 10 years of accomplishments in residential and commercial projects. Adept at estimating, purchasing, controlling materials, and executing building plans. Proven ability to select and motivate qualified workers.

Experience and Achievements

1989-present   *Construction Supervisor*
Keystone Construction Corporation, Allentown, Pennsylvania

For Penns Woods residential development of 150 homes and condominium units:

- Estimated costs, interpreted architects' plans, ordered materials, created construction schedules, hired and supervised crew of 22. Project to be completed ahead of schedule.
- Made successful presentation to lender for project refinancing, which resulted in savings of $50,000.

1987-1989   *Master Carpenter and Crew Leader*
Lehigh Valley Development, Mount Bethel, Pennsylvania

For 500,000-square-foot retail/office center in Mount Bethel:

- Directed layout, exterior fabrication, and interior installation. Included heating, ventilation & air conditioning (HVAC), plumbing, electric, wall and floor coverings.

1985-1987   *Journeyman Carpenter/Estimator*
Craft Building and Supply, Chester, Pennsylvania

Assisted master carpenters on construction or renovation of both commercial and residential structures. Learned and assumed responsibility for purchasing and estimating of jobs.

1981-1985   *Sergeant, U.S. Army Corps of Engineers*

Crew leader for various construction projects on army bases throughout the United States.

Education: **U.S. Army Engineering School, Fort Belvedere, Maryland**
Certificate in Carpentry (Journeyman Equivalent)

Great Lakes Institute, Erie, Pennsylvania
Construction Trades Certificate (one-year)

Lehigh Valley Technical High School, Mount Bethel, Pennsylvania
Classes and practical experience in carpentry, drafting, and mechanical design.

References: Provided on request, once mutual interest has been established.

WILLIAM MORGAN
853 Riverdell Road
New York, NY 10036
(212) 865-3410

OBJECTIVE: (*1*) Responsible position in Corporate Finance or Accounting offering
excellent advancement to senior management.

EDUCATION: M.B.A. Wharton School, Un. of Pennsylvania, 1979
Major: Finance   G.P.A. 3/95/4.0
President Beta Gamma Sigma Honorary

(*2*) B.A. Business Administration, Harvard University, 1977
Major Accounting   G.P.A. 3/95/4.0
Magna Cum Laude

C.P.A. May 1979

EXPERIENCE:

1981 - Present   BEATTY ELECTRONICS, INC. (CORPORATE OFFICES)
Assistant Corporate Controller
Direct Cost Accounting, Tax Accounting, and Data Processing
functions (36 employees) for this Fortune 500 manufacturer of cable
(*3*) T.V. components (annual sales $1.2 billion). Provide guidance to 6
Plant Controllers in cost accounting and tax practices. Major
accomplishments: implemented new equipment, asset valuation
program (annual tax savings $1 million), developed inventory cost
accounting system (annual savings $750,000), installation/start-up
new online, real-time order entry system.

1979-1981   BEATTY ELECTRONICS, INC. (T.V. COMPONENTS DIVISION)
Assistant Division Controller (1980-1981)
Reported to Division Vice President of Administration with full accountability for all Accounting, Financial, and Data Processing
(*4*) functions for this 1200-employee Division (annual sales $500 mil-
lion). Revamped brand manufacturing cost accounting system
resulting in substantially improved managerial cost control. Developed and installed computerized brand costing system resulting in
elimination of 12 positions and annual payroll savings of $275,000.

Brand Cost Analyst (1979-1980)
Development and maintenance of brand costing system.

PERSONAL: Age 27
(*5*) Married
U.S. Citizen

REFERENCES: Excellent references furnished upon request.

145

General Issues

Justifiably proud as he is of his sterling academic achievements, Morgan's education belongs at the end of his resume. He has excellent business accomplishments that are current. One way to make them shine out is by using a linear format.

1. Omit objective; use summary.
2. Move education to end of resume, and omit dates.
3. Edit into linear format.
4. See #3.
5. Delete personal information; U.S. citizenship might become an issue only if he is going to work for a defense contractor. Given the state of that industry, Morgan might do well to stay away.

WILLIAM MORGAN
853 Riverdell Road
New York, NY 10036
(212) 865-3410

__SUMMARY__: Fortune 500 CPA controller with billion dollar electronics manufacturer. Established asset valuation program with annual tax savings of $1MM and $750M savings through new inventory cost accounting system. Versatile team player with 15 years' experience toward senior management. Wharton and Harvard graduate.

EXPERIENCE

1981-Present   __BEATTY ELECTRONICS, INC. (CORP. OFFICES), NYC__

Assistant Corporate Controller
Report to Vice President/Finance for this Fortune 500 manufacturer of cable TV components with annual sales of $1.2 billion.
* Perform direct cost accounting, tax accounting, and data-processing guidance to six plant controllers.
* Implemented new equipment asset valuation program yielding $1MM in annual tax savings.
* Developed inventory cost accounting system with $750M savings.
* Directed installation/start-up of new online, real-time order entry system.

1979-1981   __BEATTY ELECTRONICS, INC. (TV COMPONENTS DIVISION), NYC__

Assistant Division Controller (1980-1981)
Reported to Division Vice President/Administration with full accountability for all Accounting, Financial, and Data Processing functions for this 1200 employee division with $500MM annual sales.
* Developed and installed computerized brand costing system; eliminated 12 positions at annual savings of $275M.
* Revamped brand manufacturing cost accounting system.

Brand Cost Analyst (1979-1980)

EDUCATION

C.P.A.

M.B.A. Wharton School, University of Pennsylvania
Major: Finance   G.P.A. 3.95/4.0
President Beta Gamma Sigma Honorary

B.A. Business Administration, Harvard University
Major: Accounting   G.P.A. 3.95/4.0
Magna Cum Laude

REFERENCES FURNISHED UPON REQUEST

PROBLEM RESUME: DRAFTER

RESUME

(1) Name: Randall Cardenas   **(2)** Office: 4312 W. Spruce
Residence: 9821 Fulbright St.   Omaha, Nebraska, 68113
             Omaha, Nebraska 68094   (402) 853-5321
             (402) 335-0956

DRAFTER

(3) Job Objective: To be employed as a drafter with a large
        company.

Professional Experience - Drafter - 1985 to present
        Lincoln Manufacturering Co., Omaha, Nebraska
        Duties cover completing projects on three schools, a
        library, a shopping mall, and other public sites.

Educational Background
        (4) Frontier Community College, Bellevue, Nebraska
        Enrolled in evening computer programming course (at
        present); expect AA degree in 1992
        Mechanics Institute, Lincoln, Nebraska, drafting
        curriculum 1983-1985
        (5) East High School, Omaha, Nebraska, 1979-1983

(6) Personal
        Born: May 17, 1966, in Hanson, Nebraska
        Height: 6'0
        Weight: 220
        Status: Single
        Health: Good

(7) Interests
        All sports, mountain climbing, photography. Active in
        civic and community affairs.

General Issues

The problem with this resume is that it says nothing. It is as barebones as can be. If Cardenas is to realize his goal, he has to put meat on the frame of this resume. Given his single job, his best choice is a functional resume format, after he reviews the materials in Chapter 3.

1. Incorrect positioning of identifying information.
2. Omit office number.
3. Omit "Job objective."
4. Incorrect order.
5. High school could be relevant if he took drafting courses.
6. Omit "Personal."
7. Omit "Interests."

RANDALL CARDENAS
9821 Fulbright Street
Omaha, NE 68094
(402) 335-0956

EXPERIENCE IN DRAFTING

Working Plans   Draw finished interior designs. Work on expanded views. Emphasize public works. Projects to date include three schools, a library, shopping mall, and bridge.

Final Sketches   Make final sketch of proposed drawing. Check dimensions of parts, materials to be used, and their relationship to one another.

Inking   Ink in lines and do finished lettering on blueprints.

Charts   Create finished charts based on statistical data supplied by engineering departments.

Skills   Able to visualize objectives and commit rapidly to paper. Specialize in precision work. Can translate oral instructions into desired graphic formats.

EMPLOYMENT

1985 to Present   Drafter, Lincoln Manufacturing Co., Lincoln, NE

EDUCATION

Frontier Community College, Bellevue, NE—AA degree expected in 1992 (currently completing computer programming and CAD/CAM coursework).

Mechanics Institute, Lincoln, NE—Drafting curriculum included architectural drafting, mechanical drawing, basic electronics, and industrial design.

East High School, Omaha, NE—Relevant courses included descriptive geometry, trigonometry, and physics.

COMMUNITY INVOLVEMENT

Member, Homes for Nebraskans, civic organization refurbishing homes for low-income community members.

REFERENCES AVAILABLE UPON REQUST

Richard Warren Johansen
1192 Ramsgate Court
Simi Valley, CA 93065
(805) 459-8342

① **OBJECTIVE:**

To obtain a position as Manager or Director of Environmental Services in a challenging health care setting where my commitment to top quality, cost effectiveness, and responsiveness address a high expectation.

②**BACKGROUND:**

Director of Environmental Services
③ Westridge Community Hospital
④ February 1991 to March 1993

Retained by Westridge Community Hospital after service contract ended (see FDI below). Responsible for the overall cleanliness of the hospital's environment, which includes 176 acute beds, 25 skilled nursing facility ⑤ beds, an off-site medical building, a surgical center, urgent care center, linen distribution, and grounds maintenance. I plan, organize, and distribute workload in a cost-effective manner with high standards to conform with JCAHO and other regulatory agency guidelines. Some of my achievements ⑥ have been the following: Set up a Quality Management Program that assures top quality and productivity while consistently monitoring and evaluating each employee in a quantitative measure to be used in conjunction with annual reviews. Developed a recycling program to produce revenue. Implemented waste stream standard on medical and solid waste. Oversaw landscaping renovation of the Medical Staff and Medical Records area. This included space planning, architectural design, new furniture, and area move. Successfully saw the hospital through their last two JCAHO Surveys.

⑦ **Director of Environmental Services (Contract Management)**
Food Dimensions Incorporated (FDI) at
Westridge Community Hospital
⑧ August 1988 to February 1991

Responsible for all financial systems to include monthly billing, profit and loss data, budgeting of both FDI and WCH. Trained by FDI on IBM Symphony Computer System to assist in all financial information and biweekly scheduling, cycle cleaning, duty list, and quality assurance program. Planned, organized, and distributed workload in a cost-effective manner to include the standards of FDI and Westridge Community Hospital.

Operations Manager/Assistant Director
Food Dimension Incorporated (FDI) at
Royal Oaks Hospital
⑨ August 1987 to 1988

Responsible for coordinating all three shift operations for this high-activity 277-bed hospital. Developed and implemented a criteria-based quality assurance program that supported job descriptions. Interfaced with all departments to insure communication and timely response. Assisted in staffing, hiring, training, scheduling, and counseling.

Janitorial Instructor/Coordinator
Casa Bonita Career Development Center
Vocational Workshop
⑩ August 1985 to August 1987

Developed a program to train disabled persons to seek gainful employment. Developed a curriculum, training programs, and evaluation standards to assist clients (disabled persons) in successful results. Marketed the program to rehabilitation counselors, janitorial services, and the community for on-site training and service. Responsible for profit margins, scheduling, and work evaluations.

⑪ Lucero Insulation
⑫ February 1981 to August 1985

Privately owned company contracted by Southern California Gas Company to perform energy conservation measures on commercial and residential ⑬ dwelling. This included attic insulation, weather stripping, water, electric, and gas reductions. Responsible for three crews to perform all measures in a quality manner per guidelines set by Southern California Gas Company.

⑭ J & V Building Maintenance
⑮ March 1979 to August 1986

Owner and operator of a janitorial and building maintenance company. Marketed, scheduled, and planned all daily operations of the business. Building maintenance work was performed to city and state codes. ⑯

RESUME KEY FOR RICHARD WARREN JOHANSEN: DIRECTOR, ENVIRONMENTAL SERVICES

General Issues

Johansen's biggest sin is excess verbiage and several sentences that run on forever. Searching through the voluminous first paragraph and the others that follow, you find that Johansen has fine credentials. But no one would ever bother reading them. Worse, he performs syntactical faux pas that indicate sloppiness in constructing his resume. Before redoing his resume, he had to put his thoughts into plain English.

1. "Objective" is unnecessary. Like so many others, it is full of empty words. That is why I usually recommend a concise summary, instead.
2. "Background" is incorrect terminology.
3. Location of employment is missing.
4. Omit months.
5. This paragraph is "miles" too long.
6. Run-on sentence.
7. Incorrect positioning of this job; it should be tied in with one above.
8. See #4.
9. See #4.
10. See #4.
11. Contains no job title.
12. See #4.
13. Lack of an "s" on the end of the word "dwelling" throws this whole paragraph into confusion. It sounds like he worked on only one home during his 5-year tenure!
14. See #11.
15. See #4.
16. Lack of education and reference statement leaves us hanging.

Richard Warren Johansen
1192 Ramsgate Court
Simi Valley, CA 93065
(805) 459-8342

QUALIFICATIONS: Fifteen years successful progression in environmental services management qualify me for directorship. Hospital services oversight and teaching the disabled for janitorial positions highlight my experience. Excellent interface with staff and administration in all positions.

EMPLOYMENT:

1988 - 1993   WESTRIDGE COMMUNITY HOSPITAL, Simi Valley, CA
<u>Director of Environmental Services (Employed by hospital) 1991 - 1993</u>

Responsible for overall cleanliness of total hospital environment: 176 acute beds, 25 skilled nursing facility beds, off-site medical facility, surgical center, urgent care center, laundry, and grounds. Administered to JCAHO standards. Achievements include:

* Setting up and running Employee Quality Management Program linked to annual reviews.
* Developing profitable recycling program that cut costs by 15% annually and earned revenues of $10,000 in 1993.
* Meeting deadline on landscaping renovation of Medical Staff and Records area: space planning, architectural design, new furnishings, and move-in.
* Coordinating successful results of in-facility JCAHO survey.

<u>Director of Environmental Services (Contract Management) 1988 - 1993</u>
<u>Contractor: Food Dimensions Incorporated (FDI)</u>

Directed all financial systems for both FDI and WCH: monthly billing, profit and loss data, budgeting. Used IBM Symphony Computer System for financial management plus scheduling, cycle cleaning, duty list, and quality assurance program.

1987 - 1988   ROYAL OAKS HOSPITAL, Pomona, CA
<u>Operations Manager/Assistant Director (FDI Contract Management)</u>

Coordinated staff for all three shift operations of this high-activity 277-bed hospital. Developed and implemented criteria-based quality assurance program for staff. Assisted in staffing, hiring, training, scheduling, and counseling.

1985 - 1987   CASA BONITA CAREER DEVELOPMENT CENTER, Chino, CA
<u>Janitorial Instructor/Coordinator</u>

Developed and implemented curriculum and training program to teach disabled persons work skills. Marketed program to rehabilitation counselors, janitorial services, and local governments. Achieved profit margins of up to 20%.

1985 - 1987   LUCERO INSULATION, Pacoima, CA
              <u>Foreman</u>

Managed three crews at privately owned company contracted by Southern California Gas Company to perform energy convervation measures on commercial and residential dwellings.

1979 - 1986   J & V BUILDING MAINTENANCE
              <u>Owner and Operator</u>

Marketed, scheduled, and planned all daily operations to city and state maintenance codes.

EDUCATION:

Attended Los Angeles Trade Technical College (Maintenance Management Curriculum)

REFERENCES AVAILABLE UPON REQUEST

Richard M. Austin, Jr.
2441 Lakeshore Road
Lake Forest, IL 60045
(312) 244-9152

RESUME ①

② **Objective:**   Energy Manager

Summary: Sixteen years in the energy industry, with increasing responsibility, including six years in power plant process and operations, and ten years in utility program management and marketing. Strong technical background plus in-depth, current ③ marketing experience.

Background and experience include five years of one-on-one contact with residential, commercial, and industrial utility customers, advising them on energy conservation methods, electric service requirements, and utility program application.

Proven ability to target a need and develop programs to meet that need, control, and manage all facts of program implementation, promotion, and follow-up. Highly organized, efficient, and fast-paced. Hands-on management style with excellent supervisory/team leadership skills.

Highly effective in communicating and maintaining rapport with such diverse groups as media representatives, corporate executives, public officials, government agencies, and consumers.

Excellent presentation and writing skills.

Selected Achievements: Ad Administrator of conservation programs with annual budget of $1.5 million plus, responsible for ongoing program development. Redesigned and supervised production of new brochures and media campaigns. Acquired and managed 65,000-name mailing list. Developed and implemented new incentive program for field personnel and trade allies. Supervised creation and production of all promotion pieces, developed and led introductory seminars statewide, coordinated training seminars for in-house personnel, supervised and controlled follow-up mailings, and directed ongoing communications. As a result, at 6-month point program was 100% ahead of previous year's 8-month total and on target for 12-month goals.

Established and fostered relationships with equipment manufacturers to maximize cost-effectiveness to consumer of energy-efficient equipment through utility/manufacturer cost sharing.

Conducted situation analysis as integral part of formulating first corporate marketing plan.

Presently member of corporate team evaluating options available to Consolidated Utilities for new business opportunities.

Session Chair and Speaker, "The Conservation Debate," January 1987 Demand-Side Management Seminar in Chicago, Illinois.

Member, task force for "Energy Solutions" Conference, November 1987, New Orleans, Louisiana. Session Chair and Speaker. Responsible for developing and presenting preconference workshop.

Co-chair, Customer Behavior & Preference Study commissioned by the Energy Research Institute in which conservation programs I designed are part of a central case study.

Awards & Associations: Edison Electric Institute National Writing Awards Program:
<u>1st Place</u>/Commercial Category, 1984
<u>2nd Place</u>/Industrial Category, 1983
<u>Honorable Mention</u>/Commercial Category, 1982

Illinois Energy Conservation Awards, 1986 and 1987

Chairman, Board of Directors, Lake Forest, Illinois, Chamber of Commerce, 1984-1985

Employment History: Senior Marketing Analyst, Marketing Services Department, Consolidated Utilities, 1987 to present

④ Administrator, Conservation & Load Management, Consolidated Utilities, Chicago, IL 1986-1987

Energy Consultant, Consolidated Utilities, Northeastern Region, Evanston, IL, 1980 to 1986

Nuclear Reactor Operator, U.S. Navy, 1974 to 1980

Education: University of Chicago, M.B.A. with concentration in strategic management, 1989

Northern Illinois University, B.S. in Business Administration, <u>magna cum laude</u>, 1985

U.S. Naval Electronics Nuclear and Nuclear Prototype Schools

RESUME KEY FOR RICHARD M. AUSTIN, JR.: CHRONOLOGICAL/ LINEAR, ENERGY MANAGER

General Issues

Austin's obvious weakness is that his resume is way too long. Another is that you don't know his employer until the last page. He wrote it in a functional format, no doubt because he has only one employer. The problem is he goes overboard in his power summary and other minutiae of his career. He can vastly strengthen his case by modifying the resume into a choronological format, using the bulleted linear approach. In this way, he shows the progression of his career with his single employer, so it will not be held against him.

1. No need to gild the lily. Anyone reading this document knows it's a resume.
2. Omit objective.
3. Vastly shorten his summary.
4. Move experience in chronological format near top of single page.

Richard M. Austin, Jr.
2441 Lakeshore Road
Lake Forest, IL 60045
(312) 244-9152

Summary: Sixteen years of utility management with increasing responsibility, including six years in power plant process and operations, and ten years in conservation program management and marketing. Strong technical background with current record of marketing achievement.

Experience and Accomplishments

1987 to present   *Senior Marketing Analyst*
Consolidated Utilities, Chicago, Illinois
Senior staff member in corporate marketing services department responsible for forming company's competitive response and marketing strategy.
- Conducted situation analysis for CU's first marketing plan.
- Evaluated various load management technologies (e.g., cool storage, radio control devices); implemented, managed, and controlled pilot programs based on results of study.
- Key member of corporate task forces reporting to CEO and developing new business opportunities, competitive strategies, and better utilization of resources.

1986-1987   *Program Administrator, Conservation & Load Management*
Consolidated Utilities, Chicago, Illinois
Developed, managed, and controlled nationally recognized energy conservation program resulting in 5 MW of peak reduction per year. Managed annual budget of $1.5 million.
- Created, managed, and promoted successful trade ally incentive program comprised of over 1000 distributors and manufacturers. Managed all direct mail and media advertising.
- Negotiated with contractor to provide communications, training, and fulfillment services for trade ally program and supervised all contractor activities.

1985-1986   *Energy Consultant*
Consolidated Utilities, Eastern Region, Evanston, Illinois
Account executive for large commercial/industrial customers in region. Supervised all customer service, load management, and conservation activities.
- Developed and conducted numerous technical educational programs for customers (e.g., motor and motor controls, power quality, etc.).
- Won Edison Electric Institute National Writing Awards 1982, 1983, and 1984 (Honorable Mention, 2nd place, 1st place) for articles I wrote describing conservation and load management programs I devised for customers.

Education

M.B.A University of Chicago
B.S. Business Administration, Northern Illinois University, *magna cum laude*

References

Provided upon request, once mutual interest has been established.

(1) <u>Wilson J. Everett</u>

579 Rose Avenue   INDUSTRIAL FOREMAN
Dayton, Ohio 46413
(513) 832-0937

<u>SUMMARY</u>

(2)

   INDUSTRIAL FOREMAN with twenty-five of responsible leadership
experience and with high mechanical aptitude seeks position in
production with manufacturing company. Evaluations and commendations
show outstanding record in production efficiency, leadership, safety,
and labor relations. Available because last employer, Dayton Chocolate
Company, has gone out of business. (3)

<u>EXPERIENCE</u>   <u>FOREMAN Cocoa Department, Dayton Chocolate Company</u>,
              Dayton, OH
              Once one of the ten largest manufacturers of chocolate
              products in the United States, this firm discontinued
              operations in April 1994.

1978-1994   Hired on the basis of my ability to keep intricate
              machinery functional. I supervised the work of twenty-five
              hands operating cocoa presses, pulverizers, and
          (4) separators, making fifteen varieties of cocoas and
              chocolate drinks and special runs as specified by the
              laboratory. I had full authority in hiring and in
              disciplining employees, training operators, and
              establishing hours and shifts. Fullest harmony with
              employees and had minimal difficulty with Grievance
              Committee since plant was unionized in 1980.

              <u>FOREMAN Trevelyan Separators, Inc.</u> Madison, WI
              This large midwest manufacturer is one of America's
              prinicpal makers of food processing machinery. At the time
              of my employment, it had 2000 employees.

1969-1978   Hired as a tool and die maker, promoted in two years to
              the position of foreman in machine assembly department.
          (5) Ability to suggest design modifications and deep
              production moving won me the post of assistant night
              superintendent during rush periods. Enjoyed full harmony
              with employees and had very few cases before Grievance
              Committee.

          (6) Sergeant, United States Army. Fort Belvoir, Ordnance.
              Three years of active duty.

<u>EDUCATION</u>   Graduate of St. Peter's High School, Dayton, and Dayton
          (7) Tool and Die Maker's Institute.

<u>REFERENCES</u>   Full references and copies of efficiency ratings will be
              furnished upon request.

General Issues

Everett has his materials reasonably well organized, and the chronological resume fits his continuous work progress. A linear format would make his information stand out better, his phrasing needs work, and the resume needs overall editing. Layout looks bunched and crowded.

1. Correct format needed for informational section.
2. "Years" is missing.
3. Information about employer's closing is also included within Everett's "Experience" segment.
4. Phraseology requires editing.
5. See #4.
6. Since this information does not materially relate to his work performance, it is unnecessary. He *could* use it to show an ability to command employees, but it was so long ago that it is not needed. Earlier in his career, it would have been important.
7. Education needs correct format.

WILSON J. EVERETT
579 Rose Avenue
Dayton, OH 46413
(513) 832-0937

Factory foreman with 25 years production experience employed by two of the largest firms in the specialty foods processing and equipment fields. Evaluations and commendations show continuous pattern of production efficiency, leadership, safety, and skill in labor relations. High mechanical ability.

EXPERIENCE

1978 - 1994 **DAYTON CHOCOLATE COMPANY**, Dayton, OH

Foreman:
Reported to plant manager of one of the ten largest chocolate manufacturers in the United States. Plant shut down earlier this year.
* Hired based on abilities to keep intricate machinery in top operating condition.
* Supervised work of 25 employees operating cocoa presses, pulverizers, and separators who made 15 varieties of cocoas and chocolate drinks.
* Maintained complete authority over hiring, disciplining, training, and establishing hours and shifts.
* Developed and sustained excellent working relations with employees and union representatives.

1969 - 1978 **TREVELYAN SEPARATORS, INC.,** Madison, WI

Foreman: (1971 - 1978)
Reported to plant manager for one of America's principal makers of food processing equipment.
* Promoted to assistant night superintendent during rush periods.
* Directed line employees in making design modifications.
* Executed on-time production results in working with management.

Tool & Die Maker (1969-1971)

EDUCATION

* Graduate, Dayton Tool & Die Maker's Institute, Dayton, OH

REFERENCES

Full references and copies of efficiency ratings will be furnished upon request.

Blanche Monroe
4567 Blacklawn Street
Deltona, FL 37225
(813) 483-0067

OBJECTIVE

(1) To obtain a financial management position with responsibility for supervision, operations, and profitability.

EDUCATION

(2) B.S. - Accounting - Deltona State University - 1974

(3) CAREER SUMMARY

Northview Hospital   10-90 to Present
Deltona, FL

Assistant to the Controller
Prepare monthly financial statements, account reconciliations, bank analysis, and general ledger entries.

Accountemps
Deltona, FL   6-90 to 10-90

Performed accounts receivable, accounts payable, and payroll accounting functions as well as financial statement preparation for manufacturing and insurance companies.

Southern State Savings Bank   11-79 to 6-89
Miami, FL

Responsible for general accounting procedures, daily cash management, and supervision of department personnel. Prepared the monthly and quarterly regulatory financial statements and management reports. Chaired a committee for a bankwide data processing conversion, including the evaluation of equipment, procedures, and training.

First National Bank of Miami   11-74 to 11-79
Miami, FL

Responsible for financial and management reporting as well as general ledger reconciliations. Had teller, bookkeeping, and loan department clerical experiences.

(4) SIGNIFICANT ACCOMPLISHMENTS

ACCOUNTING MANAGEMENT

Developed profit center accounting and responsibility reporting for the department heads.
Participated in the implementation of the Competitive Equality Banking Act of 1987.

EFFICIENCY/AUTOMATION

Automated the general ledger system to allow for company expansion and for product and responsibility center reporting.
Centralized office supplies for branch offices by automating the purchasing and disbursing methods.

PROCEDURE ESTABLISHMENT

Improved reconciliation procedures in the accounting department.
Implemented operating procedures for teller and branch balancing function and for ATM network.

General Issues

Because she has a number of fairly strong achievements and has some specific career goals, Monroe can benefit by converting her resume to a targeted format. Since she does not anticipate moving outside the financial services field, she can use this resume for almost any job she finds appealing.

1. Convert objective to "Job Target."
2. Reposition "Education" and remove date of degree.
3. Convert "Career Summary" into "Capabilities" and, later, into "Employment History."
4. Reposition accomplishments higher up.

BLANCHE MONROE
4567 Blacklawn Street
Deltona, FL 37225
(813) 483-0067

JOB TARGET: Financial services manager with responsibility for supervision, operations, and profitability.

CAPABILITIES:
* Prepare monthly financial statements, account reconciliations, bank analysis, and general ledger entries
* Perform accounts receivable, accounts payable, and accounting functions
* Direct general accounting procedures
* Coordinate daily cash management
* Supervise department personnel

ACCOMPLISHMENTS:

ACCOUNTING MANAGEMENT
* Developed profit center accounting and responsibility reporting to meet department head needs
* Helped implement Competitive Equality Banking Act of 1987
* Chaired comittee for bankwide data processing conversion, including evaluations of equipment, procedures, and training

EFFICIENCY/AUTOMATION
* Automated general ledger system to augment company expansion plans and permit enhanced product and responsibility center reporting
* Centralized branch office supplies by automating purchasing and disbursement

PROCEDURES ESTABLISHMENT
* Improved reconciliation procedures in the accounting department
* Implemented operating procedures for teller and branch balancing function and for ATM network

EMPLOYMENT
1990 - Present Assistant to the Controller, Northview Hospital, Deltona, FL
1990 Accountant/temporary, Accountemps, Deltona, FL
1979 - 1989 Assistant Manger, Southern State Savings Bank, Miami, FL
1974 - 1979 Bank Teller, Loan Department Clerk, First Natl. Bank of Miami

EDUCATION
BS, Accounting, Deltona State University, Deltona, FL

REFERENCES WILL BE SUPPLIED UPON REQUEST

②
① MARIA R. GUADALUPE
2465 Avenida de la Guerra
Thousand Oaks, CA 91364
(805) 496-6674

③ PROFESSIONAL OBJECTIVE

To obtain a position which will utilize my experience, my talents, and management
skills in working with people and supervision of people that will benefit and contribute
④ to the grwt and success of your industry.

SUMMARY OF QUALIFICATIONS ⑤

Qualified professional background of forteen years experience in hotel management
housekeeping. Held extremely responsible positions at major hotels in Mexico and in the
United States.

Proven management, administration, and supervision skills in staffing, training, sched-
uling/assigning duties, evaluating and supervising personnel; maintained budgets,
reports/documentation; weekly reinforcement meetings with staff.

Effective bilingual oral discussion and written instructions to staff.

⑥ EDUCATIONAL BACKGROUND

High school diploma, Mexico City. Trained in computer skills Conejo Valley Adult
School in California. Certificate of Achievement for Supervisory Skills from Ramada
Hotels; issued October 10, 1991. Studying English for the past four years at scholl and
college.   ⑦

⑧ TENURE HISTORY

Radisson Hotel Agoura Hills	ASSISTANT DIRECTOR OF HOUSEKEEPING
May 1991 - July 1993	Responsible for supervision of the housekeeping staff, recruitment, training and scheduling; maintain inventory of supplies; in charge of cleanliness and optimum conditions of facilities; coordinating hotel policies and procedures.
Ritz Hotel	EXECUTIVE DIRECTOR OF HOUSEKEEPING
Acapulco, Mexico	Duties and responsibilities as above.
1982 - 1984	Managed a staff of 40 employees including room attendants, supervisors, housemen, and gardeners; Maintained public areas.

Las Brisas Hotel Acapulco, Mexico 1979-1982	ASSISTANT DIRECTOR OF HOUSEKEEPING Duties and responsibilities as above. Las Brisas is an exclusive hotel; it has huge grounds, 250 casitas or bungalows, many with private swimming pools, and private houses to rent. Managed a staff of 55 employees including room attendants, supervisors, housemen, and gardeners. Establish good report with and hosted special guests at the hotel seeing their needs and comforts were accommodated.
Holiday Inn Hotel Acapulco, Mexico 1972 - 1979	ASSISTANT DIRECTOR OF HOUSEKEEPING Duties and responsibilities as above. Managed a staff of 30 employees, including room attendants, supervisors, housemen, and gardeners in a hotel of 165 rooms.

Additional Tenure History

(10)

Rosa Mexicano Acapulco, Mexico Arts & Crafts 1984-1988	Owner Manager Established and managed a successful business offering silver jewelry and arts and crafts of Mexico.

LANGUAGES: Bilingual/fluent Spanish

REFERENCES: Professional and personal upon request.

RESUME KEY FOR MARIA R. GUADALUPE:
DIRECTOR OF HOUSEKEEPING/HOTEL

General Issues

People from other cultures have differing ideas about resume construction. Too many opt to use exotic typefaces. Another problem that Guadalupe has is that four years are missing from her chronology. Changing the format from chronological to functional may lessen the problem, although Guadalupe will certainly have to answer potential employers' questions about the missing four years.

1. Center information block.
2. Change typeface.
3. Omit objective.
4. Misspelling.
5. Misspelling.
6. Incorrect positioning of education. Note: In *this* case, it is wise that Guadalupe retains mention of high school U.S. diploma, since employers may want to know her level of educational training.
7. Misspelling.
8. Wrong terminology for Employment History.
9. Wrong word—means *rapport*.
10. Incorrect positioning.

MARIA R. GUADALUPE
2465 Avenida de la Guerra
Thousand Oaks, CA 91364
(805) 496-6674

SUMMARY: Bilingual professional housekeeper with 14 years hotel management experience in the United States and Mexico. Consistently excellent rapport with staff, management, and guests.

STAFF SUPERVISION

Managed staffs as large as 55 employees including room attendants, supervisors, housemen, and gardeners. Thoroughly skilled in staffing, training, scheduling/assigning duties, evaluating and supervising personnel. Ran weekly reinforcement meetings with staff. Provided detailed written and oral instructions in English and Spanish, as needed.

HOTEL HOUSEKEEPING

Maintained budgets. Produced reports and documentation. Responsible for ordering supplies in coordination with purchasing departments.

EMPLOYMENT

RADISSON HOTEL, Agoura Hills, CA   Assistant Director of Housekeeping 1991 - 1993

ROSA MEXICANA, Arts & Crafts, Acapulco, Mexico   Owner/Manager 1984 - 1988

RITZ HOTEL, Acapulco, Mexico   Executive Director of Housekeeping 1982 - 1984

LAS BRISAS HOTEL, Acapulco, Mexico   Asst. Director of Housekeeping 1979 - 1982
(world famous hotel with 250 casitas or bungalows, many with private pools)

HOLIDAY INN, Acapulco, Mexico   Assistant Director of Housekeeping 1972 - 1979

LANGUAGES

Bilingual. Spanish and English. Both spoken and written.

PROFESSIONAL SKILLS

Certificate of Achievement for Supervisory Skills, Ramada Hotels, October 1991

EDUCATION

Computer and English skills, Conejo Valley Adult School, Thousand Oaks, CA
High School Graduate, Mexico City, Mexico

REFERENCES

Personal and professional provided upon request.

PROBLEM RESUME: HOTEL FRONT DESK CLERK

MANFREDO CUA
(1) 2637 Valley View Street
Lakeview Terrace, CA 92357
(818) 863-4410

WORK HISTORY

I.	COMPANY	:	HOTEL NIKKO GUAM
	TYPE OF BUSINESS	:	Hotel and Restaurant
(2)	JOB TITLE	:	Front Desk Clerk
	SUPERVISOR	:	Mr. Takuji Motoyama (3)
	SALARY	:	$7.00 HR. (4)
	DATE OF EMPLOYMENT	:	January 92 - December 92
	ADDRESS	:	268 Aurora St.,
			Dedo, Guam, USA 96912 (5)

MAIN RESPONSIBILITIES

- handle check-in and check-out of hotel guests
- handle individual and group reservations
- check rooming lists on a daily basis to determine the occupancy rate of the hotel
- handle and follow up guests' complaints by informing various hotel departments about the complaint reported
- check room status to determine what room is saleable or what room needs make-up
- making packets for group check-in (key issuing, gift certificate)
- posting miscellaneous bills
- filing of different vouchers according to room number
- compiles article pick-up, lost and found items, messages, and morning call request on a log book
- foreign exchange handling, usually Yen to dollars
- close daily sales receipts

SEMINARS ATTENDED

BASIC JAPANESE LANGUAGE, Simple Japanese Conversation, February 1992

SPIRIT OF HOSPITALITY, The Hospitality Professional, April 1992

SPIRIT OF HOSPITALITY, The Essentials of Hospitality, July 1992

EDUCATION

1986 - 1991   DE LA SALLE UNIVERSITY, 2401 Taft Ave., Manila, Philippines
Degree: Bachelor of Arts

ORGANIZATION AND MEMBERSHIP ACTIVITIES

Journal of Philosophy, contributor
De La Salle Outdoor Club, member
DLSU Taekwondo Team, member
Philippine Canine Club Inc., member
Jet Ski International Guam, member
Recycle or Die Foundation, co-founder

PERSONAL DATA

⑥ Birth Date:8-2-69   Marital Status: Single
Height: 5'6"   Weight: 120 lbs

ADDITIONAL SKILLS

Fluent in English and Tagalog
Can carry simple Japanese conversation
Type 45 wpm
Knowledge of WordPerfect 5.1 and basic computer
fundamentals

General Issues

Cua has to learn the ways of American resume writing if he expects to progress in American business. Using a functional format, Cua can play up his single year of hotel experience to greater advantage. Though he is a recent graduate, his credentials from a foreign university will not carry the weight that a U.S. degree or a prestigious European school might bring.

1. Center identification block.
2. "Work History" is in incorrect format.
3. Omit supervisor.
4. Omit salary.
5. Omit address of employer.
6. Omit personal data.

MANFREDO CUA
2637 Valley View Street
Lakeview Terrace, CA 92357
(818) 863-4410

HOTEL FRONT DESK GUEST SERVICES
* Handle check-in/check-out
* Check and reconfirm individual and group reservations
* Interface between guests with complaints and management
* Provide foreign exchange transactions, usually from Yen to dollars

CLERICAL FRONT DESK RESPONSIBILITIES
* Verify occupancy rates on a daily basis
* Interface with housekeeping staff to expedite availability of clean rooms
* Assemble guest check-in packets
* Register lost-and-found items
* Post bills
* Close out daily sales receipts

SKILLS
* <u>Languages</u>: Fluent in Tagalog and English; Adept at basic Japanese conversation
* <u>Computer</u>: Knowledge of WordPerfect 5.1 and basic computer fundamentals; Keyboarding @ 45 wpm

EDUCATION
1991 - BA, De La Salle University, Manila, Philippines
1992 - Professional Seminars on Guam: Basic Japanese Language, Two "Spirit of Hospitality" Seminars

ACHIEVEMENTS AND ACTIVITIES
* Journal of Philosophy, contributor
* Recycle or Die Foundation, co-founder

EMPLOYMENT
1992 - HOTEL NIKKO, GUAM   Hotel and Restaurant Front Desk Clerk

REFERENCES WILL BE SUPPLIED UPON REQUEST

PROBLEM RESUME: HUMAN RESOURCES/EMPLOYEE BENEFITS MANAGER

①RESUME

② John M. Worthington 53 Hornsby Grove, Albany, NY 13577
(813) 353-2431

Experience

③ **Johnson Gear Internatl.** - LABOR RELATIONS MANAGER - June 1984 -
Present
Did main work in personnel department with duties which included
the following functions: labor relations, negotiations, wage and
salary decisions, interviewing and safety considerations. Was also
responsible for management of personnel office, supervising the
department. This position was a promotion from previous position in
Purchasing Department for more responsibility.

④ **First Merchandising Corp.** - PERSONNEL ADMINISTRATOR - Feb. 1982 -
May 1984
Was responsible for all the phases of administration in the
personnel dept. Also took courses in sales and marketing while
employed here, and increased my knowledge of company methods. Heavy
phone contact and correspondence.

⑤ **Allied Pump Mfg.** - ASSISTANT PERSONNEL MANAGER - June 1979 - Jan.
1982
Held responsible position of personnel manager's assistant and
performed all of her many functions when she was absent, ill, or
away. Dealt with employees and potential employees. This was a
high-pressure position.

⑥ **Right Staff Employment, Inc.** - COUNSELOR - March 1978 - April 1979
This was a company which placed over 200 people in permanent
positions each year and also handled temporary placements. We
specialized in clerical and secretaries. Learned about hiring
practices in the metro area.

Education

⑦ B.S., Fordham University, Majored in Biology with Minor in Theater,
1977
Graduated - Matthews City High School - National Honor Society 1973
⑧ Hanson Academy - Received Special Scholastic Award in Science 1969

Personal

⑨ Married: 2 dependent children
Health: Excellent

References available on request.

RESUME KEY FOR JOHN M. WORTHINGTON: HUMAN RESOURCES/EMPLOYEE BENEFITS MANAGER

General Issues

Worthington's resume is long on basic responsibilities, short on achievements. He needs to review the steps in Chapter 3 to determine the strong points of what appears to be a progressive career, and turn his resume into a more targeted format.

1. Unnecessary verbiage.
2. Organize identifying block in better fashion.
3. Edit and strengthen with achievements.
4. See #3.
5. See #3.
6. See #3.
7. Omit.
8. Omit.
9. Omit.

JOHN M. WORTHINGTON
53 Hornsby Grove
Albany, NY 13577
(813) 353-2431

JOB TARGET: Employee benefits position in which I can use my 16 years of increasingly more responsible human and labor relations experience.

CAPABILITIES:
* Direct comprehensive employee benefits programs for over 30,000 employees.
* Negotiate contracts with insurance and other benefits personnel.
* Manage labor relations negotiations and prenegotiating planning.
* Analyze health-care programs for cost-effectiveness.
* Research both salaried and hourly benefits programs.
* Closely monitor federal, state, and local legislation to identify potential labor problems.
* Accurately and clearly communicate details of benefits programs to both employer and employees.

ACHIEVEMENTS:
* Developed benefits programs for over 250 hourly employees.
* Negotiated benefits contracts with 17% savings in premiums.
* Identified potential problems in new legislation, which avoided several potentially costly lawsuits.
* Directed research on major new medical benefits program.
* Set cost-control standards that have since been adopted throughout pump industry.
* Developed new employee benefits program kit directed to hourly employees.

EMPLOYMENT:

1984-Present	Johnson Gear International, Albany, NY, Labor Relations Manager
1982-1984	First Merchandising Corp., NYC, Personnel Administrator
1979-1982	Allied Pump Mfg., Bergen, NJ, Assistant Personnel Manager
1978-1979	Right Stuff Employment, Inc., NYC, Employment Counselor

EDUCATION:
BA, Biology, Fordham University

REFERENCES: Available on request.

PROBLEM RESUME: INDUSTRIAL ENGINEER

(1) Robert Stein
303 Anson Road
Tulsa, Oklahoma 74394
(918) 584-7890

(2) Capsule: Seven years' experience as industrial engineer specializing in financial planning and personnel management. Recently passed the examination for state license. Am seeking position with small industrial firm where I can be assured of multiple responsibilities and growth.

(3) Work Chronology: 1985 to present. A.B. Johnston, Inc. (4) Title: senior industrial engineer. Headed team of eight specializing in financial planning for the firm, which showed a regular annual profit of $10 million. By cutting costs in supplies and personnel, we were able to save the company $1.5 million last year. Part of the year was devoted to an extensive analysis of new products and directions that would be open to the company through the 1990s.

1982-1985. Stonecraft Engineering Industries, Inc. (5) Began as junior engineer working under supervision of department head. In 1983, was appointed assistant to department head. (6) In 1984, when department head retired, was appointed to replace him. I introduced the team study approach, which is still used at the company today.

(7) Education: B.S. Engineering, M.I.T., 1982. (8) Full scholarship student; also worked 30 hours a week for four years during school.

References: Furnished on request.

RESUME KEY FOR ROBERT STEIN: INDUSTRIAL ENGINEER

General Issues

Original resume is a very unstructured, unsophisticated document for someone with this man's credentials. No human resources screener would look twice. Stein needed to review his background and come up with more specifics. Linear resume better fits his technical background.

1. Incorrect positioning of contact information.
2. "Capsule" is wrong verbiage. His request for a specific kind of position should be put into cover letter.
3. "Work chronology" is wrong verbiage.
4. Missing location of employer.
5. Missing location of employer.
6. Various positions with this employer need to be delineated.
7. Education is in incorrect format.
8. Omit year of graduation.

Robert Stein
303 Anson Road
Tulsa, OK 94394
(918) 584-7890

<u>Summary of Qualifications:</u> Technically accomplished industrial engineer with specialties in financial planning and personnel management. High energy. Results oriented. Excellent reputation for bringing projects in on time and at/below budget.

EXPERIENCE

1985 -
Present

A.B. Johnston, Inc., Tulsa, OK
<u>Senior Industrial Engineer</u> Report to plant manager for this $500 million office equipment manufacturer, which regularly shows a profit of $10 million.
- Appraise and recommend organization structures and functional assignments.
- Initiate, develop, and maintain present and new systems and procedures for incentives, work analysis, estimates, costs, and methods.
- Set up and direct performance measurement systems.
  - Saved firm $1.5 million last year by cutting costs in supplies and personnel.
  - Personally changed incentive standards to one for one system on various grinding operations.
  - Analyzed new products and directions for firm growth through 1990s.

1982 -
1985

Stonecraft Engineering Industries, Inc., Oklahoma City, OK
<u>Department Supervisor,</u> Industrial Engineering Dept. (1984-1985)
Reported to general manager of this $300 million aerospace manufacturer.
- Supervised investigation of method difficulties in production areas.
- Developed and maintained long-range facility planning.
  - Introduced team study approach, still used at facility today.
<u>Assistant to Department Head</u> (1983-1984)
<u>Junior Engineer</u> (1982-1983)

EDUCATION

B.S. Engineering, M.I.T.
Full scholarship student; worked 30 hour weeks during all four years.

<u>**REFERENCES**</u>: Available on Request

PROBLEM RESUME: LABORATORY TECHNICIAN

① Shelley Ford
5037 Rose Avenue
Dubuque, IA 53092
(319) 884-5630

Summary

② Nine years of experience, five as an assistant laboratory technician working on measles vaccine, and three years as a technician working on a cancer research project. Have attended various training programs and seminars in the United States and Europe; have presented five papers at seminars.

Objective

③ Seeking position as senior technician in cancer research.

④ Employment

Chesterton Laboratories, Dubuque, Iowa.
November 1985-present.
Laboratory technician working on cancer research project under supervision of Marya Knightley, Ph.D. Was sent as company representative to four international conferences; presented papers on our cancer research project at 1987 and 1988 conferences. Three other papers were presented through my initiative at U.S. Laboratory Technicians for Cancer Research annual meetings.

Harmon-Stadley Laboratories, Inc., Chicago, Illinois.
September 1980-October 1985.
Assistant laboratory technician. Harmon-Stadley, a major drug research firm, runs a six-month training program for all new technicians, in which I was a participant. Was then assigned as laboratory technician, working under John Gedding, Ph.D., specialist in vaccine research. Worked on long-term vaccine development.

⑤ Education

Purdue University
Ph.D. 1980, oncological research
M.A. 1974, chemistry
B.A. 1972, chemistry

RESUME KEY FOR SHELLEY FORD: LABORATORY TECHNICIAN

General Issues

Resume rambles without clear focus. Need to separate experience from presentations to show depth of professional standing in her field. Strong academic background belongs closer to top, due to nature of her work. Physical layout is crowded and does not convey level of expertise.

1. Name should be in caps.
2. Summary needs strengthening.
3. Objective should be removed from this chronological format.
4. Change heading from "Employment" consistent with her background to "Professional Experience."
5. Reposition education directly under summary, and omit dates.

SHELLEY FORD
5037 Rose Avenue
Dubuque, IA 53092
(319) 884-5630

SUMMARY

Laboratory technician specializing in chemotherapy research in cancer and measles vaccine research for nine years. Worked under prominent specialists, Marya Knightley, Ph.D. in cancer project, John Gedding, Ph.D. for measles research. Presented two papers at international conferences, and three at U.S. meetings detailing findings.

EDUCATION

Purdue University
>Ph.D. Oncological Research
>M.A. Chemistry
>B.A. Chemistry

PROFESSIONAL EXPERIENCE

Chesterton Laboratories, Dubuque, IA
Senior Laboratory Technician 1985 - Present

Supervise staff of four in the collection of tissue in laboratory animals dealing with proposed new chemotherapy medications for metastasized breast and ovarian cancer. Analyze statistical results and draft technical papers summarizing findings.

Harmon-Stadley Laboratories, Inc., Chicago, IL
Laboratory Technician 1980-1985

Provided daily injections to laboratory animals to test new measles vaccine. Analyzed blood and urine samples. Interfaced with senior technicians to provide quantitative and qualitative results.

Laboratory Technician Trainee

PROFESSIONAL PRESENTATIONS

*International Conference of Cancer Researchers, 1987, 1988

*U.S. Laboratory Technicians for Cancer Research, 1991, 1992, 1993

References: Personal and professional references provided upon request.

PROBLEM RESUME: LIBRARIAN

(1) JO ELLEN RIZZOLI
6436 North Winston Place
Atlanta, GA 29473
(404) 563-9444

(2) OBJECTIVE   <u>Reference Librarian: University/College</u>

(4)

AREAS OF (3) KNOWLEDGE AND EXPERIENCE	Reference/Research	Budgets Statistics Reports Report Writing
	Book Reviews	Public Relations
	Library Administration Staff Supervision	Public Speaking
	Personnel Training, Evaluation	Community Interest
	Program Development	A-V Equipment

(5) PERSONAL   Birthdate: 9-11-52   Single
5'3"   Excellent health

EDUCATION   Rosary College, River Forest, IL
M.A. Degree in Library Science - 1980
Loyola University, Chicago, Illinois
B.S. Degree in Humanities - 1976
Major: History; Minors: Spanish, French

LANGUAGES   Spanish, read and speak with moderate
fluency; French, read.

PROFESSIONAL   American Library Association
ASSOCIATIONS   Georgia Library Association

FOREIGN   Eleven European countries, Canada, Mexico
TRAVEL

EXPERIENCE   ATLANTA PUBLIC LIBRARY
(6) 1979 to
Present

(7) (1983 to   <u>Position:</u> Branch Librarian - North Lake
Present)   View Branch

RESUME KEY FOR JO ELLEN RIZZOLI: LIBRARIAN

General Issues

Resume is extremely confusing and difficult to read. It has no cohesive theme. Columnar format makes it difficult to follow. Rizzoli is also missing key achievements and makes no use of action verbs.

1. Identifying information should be centered.
2. Objective weakens scope of future potential employment opportunities. It should be eliminated.
3. "Areas of Knowledge and Experience " category is too broad; materials lack depth of information.
4. Columnar format is confusing.
5. Personal section should be omitted completely.
6. Work experience is incomplete.
7. Incorrect format and positioning.

JO ELLEN RIZZOLI
6436 North Winston Place
Atlanta, GA 29473
(404) 563-9444

Professional Experience: Branch Librarian

* Directed book selection for the largest suburban branch in the Atlanta system

* Specialized in research covering art, architecture, business, and science

* Supervised staff of 20 full-time and part-time librarians and clerical support

* Managed personnel training and evaluation programs

* Developed special interest community programs including the "Atlanta Revisited Celebration: 125 Years of Post Civil War Growth" in 1990

* Heightened public awareness via public speaking programs and book reviews

* Maintained oversight for all branch financial matters
    - budgets
    - statistical reports

Work History:

Atlanta Public Library, Atlanta, GA - largest public library system in state
    - Branch Librarian - North Lake View Branch, 1983- Present
    - Associate Librarian - Clearwater Branch, 1979 - 1983

Professional Associations:

- American Library Association
- Georgia Library Association

Languages and Foreign Travel:

- Spanish, read and speak with moderate fluency
- French, read
- Traveled to eleven European countries, Canada, Mexico

Education:

- M.A. Library Science, Rosary College, River Forest, IL
- B.S. Humanities, Loyola University, Chicago, IL

References available upon request

STANLEY GARCIA
723 NW Avenue 34
Washington, DC 20036
(202) 463-4312

EDUCATION: Columbia University, New York, New York
Majors: Politics, Philosophy
Degree: Bachelor of Arts, 1993
Grade Point average: 3.0
Regents Scholarship recipient
Columbia University Scholarship recipient

EXPERIENCE:

(1) 7/93 - 9/93   Graduate Business Library, Columbia
University, New York
General library duties, enter new students
and books onto computer files, gave out
microfiche, and reserved materials.

(2) 9/92 - 5/93   German Department, Columbia University,
New York
Performed general office duties. Extensive
assistance by phone and in person.
Collating and proofreading. Assisted
professors gather class material.

(3) 6/92 - 9/92   Loan Collections Department, Columbia
University, New York
Initiated new filing system in this office.
Checked arrears in Bursars Department
during registration period.

(4) 9/91 - 6/92   School of Continuing Education, Columbia
University, New York
Involved in heavy public contact as well as
general clerical duties.

(5) SPECIAL
ABILITIES: Total fluency in Spanish, studying German.
Can program in BASIC. Excellent research
abilities.

(6) INTERESTS: Reading, classical music, and foreign
travel

REFERENCES: Available on request.

RESUME KEY FOR STANLEY GARCIA: MANAGEMENT TRAINEE/ NEW GRADUATE

General Issues

Garcia's is the typical story of the new graduate fresh from college, with a reasonably good grade point average. He has the strength of having gone to an excellent university, but not much employment background. The functional resume is made for people like Stanley. Employers do not expect a graduate to be flush with business experience; many prefer to mold them on the job. With his revised resume, Stanley will be ready.

1. Experience format needs revision.
2. See #1.
3. See #1.
4. See #1.
5. Abilities should come after "Education" since they are a strong selling point on this resume.
6. Omit "Interests."

STANLEY GARCIA
723 NW Avenue 34
Washington, DC 20036
(202) 463-4312

EDUCATION

BA, Columbia University, New York, NY, 1993

Majors: Politics, Philosophy, GPA 3.0

Recipient: Regents Scholarship and Columbia University Scholarship

TECHNICAL SKILLS

Computer literate. IBM and Macintosh. Can program in BASIC. Excellent research abilities.

LANGUAGE SKILLS

Complete fluency in Spanish: reading, writing, conversation. Basic knowledge of German.

EXPERIENCE

OFFICE WORK: General clerical. Heavy phones. Initiated new filing system. Collating and proofreading. Assisted professors to gather class materials.

LOAN COLLECTION: Checked arrears in Bursars Department during registration.

LIBRARY: Entered new students and books onto computer files. Delivered microfiche and reserved materials.

EMPLOYMENT

1993 - COLUMBIA UNIVERSITY: Graduate Business Library
German Department

1992 - COLUMBIA UNIVERSITY: Loan Collections Department
School of Continuing Education

REFERENCES

Available on request.

PROBLEM RESUME: MANAGEMENT TRAINEE/SOME BUSINESS EXPERIENCE

Gail M. Kowalski
(1) 4525 Meridian Street, Apt. 2
Indianapolis, IN 46206
(307) 765-9876

(2) Objective: To be hired as management trainee in a large company and eventually rise through the ranks into general management.

Education: **Purdue University Extension, Indianapolis**
(3) 22 credits toward MBA. Attended full time 1988-1989.
Now attending part-time. Scheduled completion: June 1992.

University of Notre Dame
B.S. Business Administration
Summa Cum Laude

(4) Experience: **Hoosier Hospital Products, Plainfield, Indiana**
1989-Present

Customer Service Manager responsible for customer order department staff of 15. Duties include document routing from mailroom through production to shipping. Coordinate inventory needs with production and materials management department.

Accounting Department
Indiana Aluminum Corporation
North Webster, IN 46555
1984-88

Part-time job while in college. Began as part-time accounts receivable clerk (1984); promoted to Supervisor, Accounts Receivable, 1986-88. Supervised night-shift staff of three clerks. Suggested computerized accounts receivable reporting mechanism; helped improve collection of delinquent accounts.

References: Furnished upon request.

RESUME KEY FOR GAIL M. KOWALSKI: MANAGEMENT TRAINEE/ SOME BUSINESS EXPERIENCE

General Issues

Kowalski worked her way through college. After graduation, she took graduate level courses before beginning her professional job search. This resume does not reflect her business acumen, and would just be passed over. She needs a more promotional approach with more active verbs and bold statements of achievements.

1. Move identifying block to center.
2. Change "Objective" into proactive "Summary."
3. Move "Education" to bottom of resume.
4. Add "Achievements" to "Experience" and capitalize on them.

Gail M. Kowalski
4525 Meridian Street, Apt. 2
Indianapolis, IN 46206
(307) 765-9876

Summary: *Management Trainee* with B.S. in Business Administration (*summa cum laude*) and 22 graduate credits toward M.B.A. Six years of professional management with increasing responsibility in full-time, part-time, and summer positions while attending college.

Experience and Achievements

1989-present   *Customer Service Manager*
Hoosier Hospital Products
Plainfield, Indiana

Plan and direct the activities of Customer Order Department staff of 15. Manage document routing from mail room through production to shipping. Plan inventory needs with production and materials management departments.

- Initiated project with manufacturing and purchasing to create more accurate finished goods reporting, which resulted in savings of $25,000 annually and significantly reduced order fulfillment time.

1984-1988   *Supervisor, Accounts Receivable*
Indiana Aluminum Corporation
North Webster, Indiana

Began as part-time accounts receivable clerk, 1984. Promoted 1986 to supervisor of accounts receivable (night shift) in this manufacturing company with customer orders of $22 million per year.

- Developed computerized accounts receivable reporting mechanism; devised a system that reduced average collection time from 65 days to 42 days. Received bonus and commendation from corporate office as a result.

Education: Purdue University Extension, Indianapolis
M.B.A. program, 1989—full-time; 1989-present—part-time
Scheduled completion: June 1992

University of Notre Dame
B. S. Business Administration (summa cum laude)
* Maintained 3.9 GPA all four years while working an average of 30 hours per week.

References: Furnished upon request.

HENRY SILVERTON
2634 Washburn Street
Vancouver, WA 98357
(405) 801-4290

(1) JOB INTERESTS: Manufacturing Manager, Materials Management, Manufacturing Engineering.

(2) OBJECTIVE: To make a meaningful contribution to an organization by applying my knowledge of manufacturing and production processes. My enjoyment in a job comes from seeing the team I work with excel beyond their expectations.

(3) SKILLS:
Production and quality processes, Production, Planning, Materials Management, MRP II, planning systems. Created a number of JIT OPERATIONS. Established the manufacturing operations for three start-up divisions. Operations Turn Around. Procurement process and management. Team building, people involvement, coach Production problem solving. Use of personal computers.

BUSINESS EXPERIENCE

Oct. 1992 - Oct. 1993 NELSON-TILLIS CORPORATION, Tucson, AZ
Vice President of Operations
Responsible for the manufacturing operations in Tucson and Scotland, which included wafer fabrication, product assembly and test, materials planning and management, coordination of offshore assembly and process engineering. Created (4) a planning system and re-established the MRP system. Made a 15-point improvement in our "on-time" delivery performance to a new test facility that handles about 50% of our volume. Effectively loaded an offshore plastic assembly house in Taiwan. Rebuilt the direct work force and some of the supporting structure. Prepared ourselves and were certified for ISO9001 in August 1993.

January 1991-Oct. 1992 SYNTRONICS, INC., Beaverton, OR
Systems Group Manufacturing Manager and Procurement Manager
Coordination of manufacturing operations for the three divisions of the Systems (5) Group. After five months, the Systems Group was merged into the Test and Measurement Group, so I chose to become the Procurement Manager at Terminals Division. Worked with a Taiwanese OEM to bring into production a low-end X-Terminal Product Line. Worked with our Japanese OEM and division engineering on a high-resolution, low-cost, integrated X-terminal.

July 1990 - 1991
MULTIPLE INTEGRATED TECHNOLOGIES, Beaverton, OR
Vice President of Operations

(6) Responsible for the fabrication, assembly and test, processing engineering, and material support of our IC operations. End-to-end yields increased from three to six times. Output increased and we made gradual improvement in meeting customer needs and provided training in statistical problem solving. MIT was profitable for the first time in 1990.

Nov. 1986 - July 1990
SYNTRONICS, INC., Vancouver, WA
Division General Manager

(7) Responsible for engineering, marketing, manufacturing, finance, and human resources over three divisions—Integrated Systems Divisions (Products for Systems Business), Lab Instruments Division (High-End Oscilloscopes), and Accessories and Components Division (Probes, Accessories, Hybrid Assemblies, and Precision Plated Parts).
Revenue decline turned around and started to increase. Restructured manufacturing and reduced space, one division by one half. Introduced several major products in less than one year. Reduced Division people count by 33% without a layoff. Divisions turned profitable.

July 1983 - Nov. 1986
SYNTRONICS, INC., Vancouver, WA
Division Group Manufacturing Manager

(8) Responsible for the board assembly, final assembly, and test, materials operation, and manufacturing and process engineering for the Portable Oscilloscope Division Group. Reduce inventory and manufacturing cycle times by 60%. Converted to a commercial MRP system. Improved quality and reduced scrap and rework. Reduced manufacturing overhead structure. Established an environment for people involvement.

Feb. 1962 - July 1983
HEWLETT-PACKARD COMPANY, Multiple locations in United States
Manufacturing Manager
Responsible for manufacturing in Loveland and Fort Collins, Colorado, and three start-up divisions located in Brazil, Vancouver, Washington, and Mexico. Products ranged from test and measurement equipment, hand-held calculators, medical equipment printers, business computers. In Vancouver, we started the initial leading to "Just In Time" manufacturing. Also, served as Division Finance Manager for one and one half years prior to going to Brazil.

EDUCATION

(9) Aug. 1958 University of Wisconsin, Madison, Wisconsin
Bachelor of Science - Electronics Engineering
Additional Courses: Instructor Course, ISO 9000 Audit Course, Shainin Statistical Engineering Course, Mini Tech and H-P internal courses.

RESUME KEY FOR HENRY SILVERTON:
MANUFACTURING MANAGER

General Issues

Silverton is a classic example of one of the problem people in today's economy. He's an engineering type who spent many years with one firm, only to be cut loose after 20+ years of service. Since then, he's bounced around to three other employers, one of them twice. What will help Silverton?—cleaning up his resume and highlighting his impressive achievements. Lopping off the first job with its 20-year history would also make him look much younger on paper. Since he'd have to explain his considerably older appearance once he got to the prospective employer, he argues he'd rather leave it in and show employers the full range of his prowess. Finally, he should use that list of skills as keywords and enter his resume into an online resume service.

1. Merge into summary; use also in keyword listings.
2. Omit objective.
3. Use as basis of keyword list.
4. Highlight accomplishments.
5. See #4.
6. See #4.
7. Combine with earlier position, since it was the same employer.
8. See #7.
9. Omit graduation date.

HENRY SILVERTON
2634 Washburn Street
Vancouver, WA 98357
(405) 801-4290

KEYWORDS

Manufacturing Manager. Materials Management. Manufacturing Engineering. Production and Quality Processes. Production Planning. MRP II. Planning Systems. Just-In-Time, JIT Operations. Operations Turn Around. Procurement Process. Team Building. People Involvement. Coach. Production Problem Solving. Personal Computer Use.

SUMMARY: Vice president and senior manufacturing manager responsible for numerous JIT operations and establishing operations for three start-up divisions. Full involvement in all facets of materials management, engineering, and production in a high-tech environment. Team player with strong motivational skills. Undaunted by long hours.

PROFESSIONAL EXPERIENCE

NELSON-TILLIS CORPORATION, Tucson, AZ   1992 - 1993
Vice President of Operations

Maintained overall responsibility for manufacturing operations in Tucson and Scotland including wafer fabrication, product assembly and test, materials planning and management. Coordination of offshore assembly and process engineering.
Results:
* Created planning system and re-established MRP SYSTEM.
* Made 15-point improvement in "on-time" delivery to new test facility that handles 50% of firm's volume.
* Rebuilt direct work force and supporting structure.
* Guided certification for ISO9001 in August 1993.

SYNTRONICS, INC., Beaverton, OR   1991 - 1992
Systems Group Manufacturing Manager and Procurement Manager

Coordinated manufacturing operations for all three divisions of Systems Group. Became procurement manager at Terminals Division when Systems Group merged into Test and Measurement Group.
Results:
* Achieved production startup of low-end X-Terminal Product Line working with Taiwanese OEM.
* Developed high-resolution, low-cost, integrated X-terminal in tandem with Japanese OEM and division engineering.

MULTIPLE INTEGRATED TECHNOLOGIES, Beaverton, OR   1990-1991
Vice President/Operations

Directed fabrication, assembly and test, processing engineering, and materials support for IC operations.

Results:
* Increased end-to-end yields from 3 to 6 times.
* Turned corner toward profitability for first time in 1990.

SYNTRONICS, INC., Vancouver, WA   1983 - 1990
Division General Manager (1986 - 1990)

Directed enginering, marketing, manufacturing, finance, and human resources for three divisions: Integrated Systems, Lab Instruments (high-end oscilloscopes), and Accessories and Components (probes, accessories, hybrid assemblies, precision plated parts).

Results:
* Reduced divison work force by 33% without layoffs.
* Introduced several new products in less than one year.
* Helped reverse revenue decline into profitability.

Division Group Manufacturing Manager   1983 - 1986

Managed board assembly, final assembly, and test plus materials operation and manufacturing and processing engineering for Portable Oscilloscope Division Group.

Results:
* Reduced inventory and manufacturing cycle times by 60%.
* Converted to a commercial MRP system.
* Improved quality and reduced scrap and rework.
* Reduced manufacturing overhead structure.
* Introduced employee participation environment.

HEWLETT-PACKARD COMPANY, Multiple locations in U.S.   1962 - 1983
Manufacturing Manager/Division Finance Manager

Directed three start-up divisions in Brazil, Vancouver, WA, and Mexico. Ran facilities in Loveland and Fort Collins, CO. Products included test and measurement equipment, hand-held caculators, medical equipment printers, business computers.

Results: * Initiated initial concept leading to "Just In Time" manufacturing.

EDUCATION

BSEE, University of Wisconsin, Madison, WI

Professional Coursework: Instructor Course, ISO 9000 Audit Course, Shainin Statistical Engineering Course, Mini Tech and H-P internal courses.

REFERENCES AVAILABLE UPON REQUEST.

PROBLEM RESUME: MARKETING MANAGER

(1) NORMAN WILDER 4235 Seacrest Avenue, Canoga Park, CA 91306 (818) 346-6533

(2) **OBJECTIVE:** A management position encompassing design/planning, development, presentation, and marketing.

QUALIFICATIONS: Fifteen years of diverse design, development, and marketing experience including: project management and supervision; sales and marketing; logistical evaluation and revision; budgeting; employee recruiting, selection, and evaluation; conducting employee and staff training sessions; retail management. Strong planning, creative, and presentation skills.

(3) **ACHIEVEMENTS:** **Prepared** a land use study and environmental impact report for a proposed public park area. The report was used in a presentation to the state for land use consideration and projected funding.

Designed and prepared landscape architecture plans for the Los Angeles City Department of Recreation and Parks, including: a city hospital; civic center; low income housing project recreation facility.

Managed landscape architecture projects from design team supervision, client and city presentations, through postconstruction inspections.

Negotiated with theater exhibitors regarding duration of film run in primary theaters, and later secondary theaters.

Tracked competitors' films in order for company to determine when and where to market Orion films.

Reduced receivables by negotiating for and receiving payments of back fees owed to company by out-of-state theater owners.

Saved time and money by implementing changes in production operation through evaluation of project logistics.

Recruited, hired, and trained new employees for business and retail.

(4) **EXPERIENCE:**

Landscape Designer and Construction Supervisor	Self-employed
Landscape Architecture Project Manager	Calvin Abe Associates
	Raymond Hansen Associates
Staff Landscape Architect	Lee Newman Associates
Sales and Marketing	Orion Pictures, Los Angeles Branch Office
	American International Pictures
Operations Manager	Resources Development Corporation
Retail Manager	Leo's Stereo

EDUCATION:

Certification - Landscape Architecture	UCLA
Bachelor of Arts - Music	USC
(5) Continuing Education	UCLA Extension

197

General Issues

Like others who have followed one career path and then abruptly shifted, Wilder's resume looks schizophrenic. Attempting to lump all his experience into the "Achievements" category is incorrect and confusing. A better solution is to opt for a functional resume and list related experience together. If he plans to continue in the landscape field, a targeted resume minus his film background would be the way to go.

1. Information strung out along the top of the page is incorrect.
2. Omit objective. Especially in his case, it says nothing.
3. "Achievements" heading does not properly reflect information in this section.
4. He has too many employers. Selective pruning would be advisable. He also does not list dates. While his reasons are understandable, they are also transparent to a human resources pro. Let's hope he finds a smaller firm that is more flexible in its adherence to a formal career path.
5. Omit "Continuing Education." It says nothing specific about courses pursued regarding his professional goals.

NORMAN WILDER
4235 Seacrest Avenue
Canoga Park, CA 91306
(818) 346-6533

SUMMARY Manager with 15 years diverse design, development, and marketing experience including: project management and supervision; sales and market planning; logistical evaluation and revision; budgeting; employee recruiting, selection, and evaluation; and conducting staff training sessions. Strong planning, creative, and presentation skills.

EXPERIENCE

LANDSCAPE PROFESSION

* Prepared park lands study and Environmental Impact Report presented to state for land use approval and funding purposes.

* Designed and prepared landscape architecture plans to City of Los Angeles Department of Recreation and Parks for: a city hospital, civic center, and low income housing project recreation facility.

* Managed landscape architecture projects: from design team supervision, to client and city negotiations, through postconstruction inspection.

FILM INDUSTRY

* Negotiated with theater exhibitors for film runs in primary and secondary theaters.

* Marketed older films for special screenings in holiday packages and film festivals.

* Tracked competition to determine best strategies for marketing Orion films.

* Negotiated overdue payments of back fees owed to Orion by out-of-state theater owners.

* Evaluated production logistics, resulting in time and money savings in production.

EMPLOYMENT

Landscape Designer and Construction Supervisor   Self-Employed
Landscape Architecture Project Manager   Calvin Abe Associates
                                                    Raymond Hansen Associates
Staff Landscape Architect   Lee Newman Associates
Sales and Marketing Manager   Orion Pictures, Los Angeles Branch Office
                                          American International Pictures

EDUCATION

Certification - Landscape Architecture   UCLA
Bachelor of Arts - Music   USC

NORMAN WILDER
4235 Seacrest Avenue
Canoga Park, CA 91306
(818) 346-6533

JOB TARGET: Landscape management position in which I can use my 15 years
design, development, and marketing experience.

CAPABILITIES:

* Project management and supervision

* Sales and market planning

* Logistical evaluation and revision

* Budgeting

* Employee recruiting, selection, and evaluation

* Conducting staff training sessions

* Strong planning, creative, and presentation skills

ACHIEVEMENTS:

* Prepared park lands study and Environmental Impact Report presented to
state for land use approval and funding purposes.

* Designed and prepared landscape architecture plans to City of Los Angeles
Department of Recreation and Parks for: a city hospital, civic center, and
low income housing project recreation facility.

* Managed landscape architecture projects: from design team supervision, to
client and city negotiations, through postconstruction inspection.

EMPLOYMENT:

Landscape Designer and Construction Supervisor   Self-Employed

Landscape Architecture Project Manager   Calvin Abe Associates
Raymond Hansen Associates

Staff Landscape Architect   Lee Newman Associates

EDUCATION:

Certification - Landscape Architecture   UCLA
Bachelor of Arts - Music   USC

REFERENCES WILL BE SUPPLIED UPON REQUEST.

PROBLEM RESUME: MARKET RESEARCH & ANALYSIS/ NEW GRADUATE

Lawrence E. Hill

LOCAL ADDRESS	PERMANENT ADDRESS
104 Johnson Walk	1953 Rt. 243
Bryan, TX 77801	Spring Valley, OH 45370
(409) 346-8225	(513) 784-3184 ①

EDUCATION

② Texas A&M University, Masters in Business Administration, expected May 1994
Overall GPA 3.7

③ Wright State University, Dayton, OH, BS of Finance, BS of Marketing
June 1992; Overall GPA 3.45

④ EXPERIENCE

Graduate Assistant, Texas A&M University, Accounting Department, College Station, TX
Aid two professors with research and the subsequent inputing of data into computer spreadsheets for analysis. Proofreading of textbooks for national publication and grading and administering of exams (Aug. '92 to present).

⑤ Computing Supervisor, Wright State University, University Writing Center, Dayton, OH
Supervised and consulted with network supervisors on the use and upkeep of the UWC'S Local Area Network; helped students and faculty to become proficient computer users; conducted brief training seminars (Sept. '89 to June '92).

⑥ English Tutor, Wright State University, University Writing Center, Dayton, OH
Worked with students of various skill levels in revising and editing their papers, ranging from simple essays to research papers. Selected to aid certain professors in working in a special computer-oriented teaching environment (Sept. '89 to Mar. '92).

⑦ Expediter, Applebee's Neighborhood Grill & Bar, Miamisburg, OH
Coordinated activities between the servers and the kitchen staff. Assisted managers in ensuring and maintaining quality levels. Helped to prepare management trainees in this area of restaurant operations. Also worked as a server during this period (Feb. '90 to Aug. '92).

Supervisor, Friendly's Restaurant, Beavercreek, OH
Oversaw the management of the restaurant in the general manager's place. Responsible for scheduling, bank deposits, light bookkeeping, and overall operation of the restaurant. Also worked as a server and trainer (Oct. '87 to Jan. '90).

⑧ HONORS AND ACTIVITIES

Academic Scholarship, Texas A & M University

Graduate Assistant to Drs. Helen Powell and Meyer Lissman, professors of Accounting, TAMU

Vice President, Public Relations & Marketing, Graduate International Business Society, TAMU

1994 Case Competition (prepared actual business strategies for an outside firm), TAMU

Exchange Student, Okayama, Japan, Summer 1990 (scholarship recipient)

Jr. Board Member, Riverbend Arts Center, Dayton, Ohio

Raised funding for the Special Wish Foundation; member of the local chapter

Academic Scholarship, Wright State University

"Outstanding" recognition in advertising competition, Wright State University

⑨ COMPUTER SKILLS

Lotus 1-2-3, Lotus Improv, WordPerfect, Microsoft Word, Microsoft Windows, Harvard Graphics, PowerPoint, dBASE, Paradox, Norton Textra, Excel, CorelDRAW!, ProComm, Turbo Pascal. I have some experience with both the Apple Macintosh and the Internet computer system as well.

RESUME KEY FOR LAWRENCE E. HILL: MARKET RESEARCH & ANALYSIS/NEW GRADUATE

General Issues

Though academically advanced, Hill is a novice to the world of resume writing. Like so many other graduates who have seen resumes of those further into their careers, he makes the mistake of putting too much emphasis on his employment record. What counts at this point are his academic credentials. Unfortunately, he's also positioned his outstanding activities roster and computer skills so far toward the end of his resume, they might never get noticed. With all those overlapping part-time jobs, Hill's far better off with a functional resume.

1. Graphic line element is unnecessary and distracting.
2. Needs to reverse order of degree and university.
3. See #2.
4. Experience is in wrong place.
5. Multiple job overlap is very confusing.
6. See #5.
7. See #5.
8. This section belongs directly below "Education."
9. This section belongs directly below "Honors and Activities."

LAWRENCE E. HILL

[University Address]
104 Johnson Walk
Bryan, TX 77801
(409) 346-8225

[Permanent Address]
1953 Rt. 243
Spring Valley, OH 45370
(513) 784-3184

EDUCATION

1994 Anticipated MBA, Texas A & M University, Bryan, TX, Overall GPA 3.7

1992 BS, Finance, Wright State University, Dayton, OH , Overall GPA 3.45

UNIVERSITY HONORS & ACTIVITIES

At Texas A & M University:

* Vice President, Public Relations & Marketing, Graduate Intl. Business Soc.
* 1994 Case Competition (Created business strategies for outside business)
* Graduate Asst. to Helen Powell, Ph.D. and Meyer Lissman, Ph.D. Acctg. professors
* Academic Scholarship

At Wright State University:

* "Outstanding" recognition, advertising competition
* Junior board member, Riverbend Arts Center, Dayton
* Member and fund raiser, "Special Wish"
* Academic Scholarship

COMPUTER SKILLS

Lotus 1-2-3, Lotus Improv, WordPerfect, Microsoft Word, Microsoft Windows, Harvard Graphics, PowerPoint, dBASE, Paradox, Norton Textra, Excel, CorelDRAW!, ProComm, Turbo Pascal. Experience with Apple Macintosh and Internet system.

EXPERIENCE

Computer Consulting: Supervised and consulted with network supervisors re upkeep and use of Local Area Networks. Conducted training seminars for instructors and students to advance their computer skills. Worked with professors in special computer-oriented teaching environment.

Research: Accounting department studies and inputing of data into computer spreadsheets for analysis.

Teaching: Graded and administered exams. Worked with students of various skill levels in revising and editing papers.

Management: Responsible for scheduling, bank deposits, light bookkeeping at family-style restaurant. Coordinated activities between servers and kitchen staff. Coordinated with managers to maintain quality.

REFERENCES WILL BE SUPPLIED UPON REQUEST

NANCY C. CAREY
15 Hammersmith Road
Philadelphia, PA 19184
(215) 542-7421

MEDICAL RECORDS ADMINISTRATOR

POSITION OBJECTIVE:

①

A challenging position in the medical information or
medical records field where extensive experience and
education may be utilized.

SUMMARY OF QUALIFICATIONS:

Diversified practical experience in the following
disciplines: Research compilation of scientific/medical
② data...Abstracting and indexing medical and scientific
documents...Maintaining and updating storage and retrieval
systems...Clinical laboratory techniques...Quality control
methods and procedures.

Demonstrated ability to accept increasing
responsibility...Capable of initiating and implementing major
decisions.

Leadership qualities combined with ability to pioneer
in new techniques and projects...Ability to communicate
effectively at all levels...Experienced in applying
principles of good management.

EMPLOYMENT RECORD:

③

October 1987 - Present   ④ Stewart Pharmaceutical Corp.
                                  1973 Fallbrook Street
Assistant Medical Librarian   Philadelphia, PA 18436

Responsibilities include abstracting and indexing articles
⑤ from medical journals...searching and compiling literature
for various company divisions, such as sales, legal, and
marketing...setting up and maintaining proper storage and
retrieval systems...cross-referencing abstract cards according
to subject heading and subheading...retrieving all published
data required for submission to the F.D.A.

<u>January 1984-September 1987</u>   Lyons General Hospital
Raleigh, NC

Chief Technician

(6) Responsible for the organization of a functional clinical laboratory in setting up and standardizing clinical procedures and practices...procurement of supplies and equipment for testing and analysis of laboratory specimens.

<u>December 1979 to January 1984</u>   Foster Medical Center
Raleigh, NC

Clinical Technician

(7) Performed routine clinical testing and analysis of specimens submitted from various departments of the hospital. Assisted chief laboratory technician in all administrative functions of the department. Collected data and made statistical analysis of all tests that funnelled through the laboratory. Submitted data to department heads when requested.

Summer and part-time jobs while attending college.

<u>April 1974 to September 1977</u>   National Hospital
Raleigh, NC

Laboratory Intern

(8) Participated in laboratory methodology and principles. Primary function was private testing for director of laboratory.

EDUCATION:

(9) Medical Laboratory Technology, Smith Community College, Raleigh, NC 1972
B.S. Biology - Duke University, Raleigh, NC 1973-1977

---

(10) PERSONAL DATA:

License: American Medical Technologist Assn.

General Issues

Though she has an excellent background, the format of Carey's resume precludes giving it more than a passing glance by a busy human resources staffer. The text is too strung out along the page, there are not enough graphic eye-catchers in the typeface, and the horizontal lines are a distraction. Carey is trying to communicate too much and has let the resume run too long.

1. Omit "Position Objective."
2. "Summary of Qualifications" needs to be substantially shortened.
3. Months should be omitted from employment data.
4. Address should be omitted.
5. Verbiage needs to be pared down, with stronger verbs used.
6. See #5.
7. See #5.
8. Unnecessary to use part-time job with such strong later full-time positions.
9. Transpose positioning of community college information. Ordinarily, it would not be necessary to keep it. Given her field of endeavor, it shows an important transition, and lends strength to credentials.
10. "Personal Data" is incorrect heading.

NANCY C. CAREY
15 Hammersmith Road
Philadelphia, PA 19184
(215) 542-7421

Summary: Professional medical records administrator with proven leadership capabilities; consistent manager with ability to communicate effectively at all levels and pioneer new techniques and methods. Over 15 years of continuously increasing responsibilities.

EMPLOYMENT

1987 - Present   Assistant Medical Librarian
Stewart Pharmaceutical Corp., Philadelphia, PA
* Set up corporate storage and retrieval systems.
* Abstract and index medical journal articles for major corporate divisions, i.e., sales, legal, marketing.
* Coordinate published data for submission to F.D.A.
* Cross-reference abstracts for ease of retrieval.

1984 - 1987   Chief Technician
Lyons General Hospital, Raleigh, NC
* Set up and standardized procedures and practices for clinical laboratory.
* Directed procurement for all supplies and equipment used in specimen testing.

1979 - 1984   Clinical Technician
Foster Medical Center, Raleigh, NC
* Tested and analyzed specimens from departments throughout facility.
* Interfaced with department heads regarding data.
* Collected raw data and performed statistical analysis of all departmental tests.
* Assisted chief laboratory technician with administrative functions.

CERTIFICATION
Licensed by American Medical Technologists Association.

EDUCATION
B.S., Biology - Duke University, Raleigh, NC
Medical Laboratory Technology, Smith Community College, Raleigh, NC

REFERENCES
Provided upon request

(1)
<u>RESUME</u>

Mark Langley
23 Bayside Shores Drive
Charleston, South Carolina 23717
(813) 345-0444

<u>WORK HISTORY</u>

(2) Strong sales background with ability to teach and transfer; excellent letter and copy techniques; capability of creating, preparing, and articulating complete original presentation; quality of addressing fresh or unusual problems and making opportunities of them.

(3) <u>NOVEMBER 1982 TO PRESENT:</u> Classified Manager, <u>The Charlestonian</u>, Charleston, South Carolina. Morning circulation 60,000 daily and 70,000 Sunday. Set up new system for Classified Outside Sales and created phone sales department. Entire new format for section. A broad incentive plan for personnel and creative rates for business contracts and nonbusiness (person to person). Revenue has more than doubled in less than two years with projected Classified income well over $7,000,000 in 1994.

(4) <u>SEPTEMBER 1979 TO OCTOBER 1982:</u> Advertising Manager, <u>Times Star</u>, Raleigh, North Carolina. Evening circulation 30,000. Involved in all facets of change from hot metal to cold type plus computer startup and training personnel on use of VDT for ads. Reduced large credit backlog with almost no customer loss and held modest gain in heavy economically depressed area. Set up total market coverage with 60,000 once-a-week inserts.

(5) <u>JANUARY 1975 TO SEPTEMBER 1979:</u> <u>The Raleigh Evening and Sunday Chronicle</u>, Raleigh, North Carolina. Served as salesman, outside sales supervisor, outside training manager, and Phone Advertising Manager. As salesman doubled percentage of field in most assigned areas and was very successful in new sales areas. For several years was responsible for the hiring and training of Classified and Retail ad sales personnel. Was first to use Wats line in phone room and was responsible for computer changeover as Phone Ad Manager. Covered major companies and North Carolina/South Carolina ad agencies handling recruitment and institutional advertising.

(6) <u>PRIOR:</u> Yellow Pages sales; previous time with <u>Chronicle</u> as well as personal ad agency work on West Coast and in Las Vegas.

<u>BUSINESS AWARDS</u>: Raleigh Chronicle "SALESMAN OF THE YEAR, 1977"
North Carolina Association of Personnel Services
"MAN OF THE YEAR" 1979

PERSONAL HISTORY

EDUCATION: North Carolina State Night Program in Marketing
               North Carolina State Management School

(7) MILITARY: U.S. Army and Security Agency and Special Services with
               two-year overseas tour in Germany.

(8) HEALTH: Good

RESUME KEY FOR MARK LANGLEY: ADVERTISING MANAGER/CLASSIFIED

General Issues

Langley makes the mistake of confusing quantity with quality. His background is excellent; he simply needs to edit his materials and omit some extraneous categories like "Military" and "Health."

1. Omit "Resume." It is obvious.
2. Tighten "Work History" and reformat into "Summary."
3. Edit and tighten.
4. See #3.
5. See #3.
6. Omit.
7 Omit.
8. Omit.

Mark Langley
23 Bayside Shores Drive
Charleston, South Carolina 23717
(813) 345-0444

SUMMARY: Classified advertising manager with nearly 20 years experience in major southeastern markets. Strong sales background. Award winner capable of creating, preparing, and articulating complete original presentations and solutions to problems.

EMPLOYMENT

1982 - Present   **The Charlestonian, Charleston, SC**
Classified Manager
Paper has a daily circulation of 60,000 and a Sunday circulation of 70,000
* Created new system for Classified Outside Sales and set up phone sales department
* Successfully implemented new formation for classified section
* Established broad incentive plan for personnel
* Developed creative rates for business contracts and nonbusiness advertising
* Doubled revenues in less than two years; projected classified income for 1994: $7,000,000

1979-1982   **Times Star, Raleigh, NC**
Advertising Manager
* Involved in all facets of change from hot metal to cold type
* Trained personnel on use of VDT for ads
* Reduced large credit backlog with almost no customer loss; held modest gain in heavily depressed economic region
* Set up total market coverage with 60,000 once-a-week inserts

1975-1979   **Evening and Sunday Chronicle, Raleigh, NC**
Sales Supervisor
* Served as outside salesman
* Managed phone advertising
* Hired and trained all new personnel in Classified and Retail ad sales
* Covered major companies and Carolina ad agencies scheduling recruitment and institutional advertising

BUSINESS AWARDS

1979   Voted "Man of the Year" by the North Carolina Association of Personnel Services, Raleigh, NC

1977   "Salesman of the Year" Raleigh Chronicle

EDUCATION

North Carolina State Night Program in Marketing
North Carolina State Management School

REFERENCES Available on request

Morgan Irving ①
10345 Warrensgate Road
Madison, WI 53723
(608) 445-0932

② Objective: Researcher/Staff Writer in news department on a newspaper or magazine

③ Experience:

* Wrote regular column on political issues for <u>University News</u>: interviewed people in the news, won community service award for ongoing series "National Economics on the Campus"

  1990-1993

* Statistician-writer for "Citizen Alert" pamphlets on crime, pollution, neighborhood renovation, and legal aid services (sponsored by the Green Bay City Council)

  Summer 1991

* Researcher and assistant speech writer for Congressman Paul Ross, Legislative Assembly Internship, Madison, WI

  Feb-April 1991

* Assistant researcher and statistician for CBS opinion poll. Election 1992. Regional Election Headquarters, Milwaukee, WI

  Fall 1992

④ Education:

1993 B.A. in Political Science (minor in journalism)
University of Wisconsin, Madison

⑤ Other Work Experience:

waiter, camp counselor, shoe salesman

References and writing samples available on request.

RESUME KEY FOR MORGAN IRVING: NEWSPAPER WRITER/ NEW GRADUATE

General Issues

Lacks focus. Resume format wrong for new graduate. Needs stronger emphasis on education near the top. With his desire to target a specific job, he overlooked the need to explain his capabilities.

1. Name should be in caps.
2. Remove "Objective"; replace with term "Job Target."
3. Move bulk of "Experience" into "Achievements."
4. Education belongs closer to top. It needs to be expanded with coursework. "Legislative Internship" belongs under education, not "Experience."
5. "Other work experience" is so far afield, it is irrelevant.

MORGAN IRVING
10345 Warrensgate Road
Madison, WI 53723
(608) 445-0932

Job Target: Researcher/Staff Writer in news department on a newspaper or magazine

EDUCATION:

B.A. Political Science with minor in Journalism
1993 University of Wisconsin, Madison

Related Coursework:
* Legislative Functions at State and National Levels
* Economics & Its Role In 20th Century Government Models
* Journalism: The "Free Press" from 1970 Onward
* Advanced Editing
* Writing for the Op/Ed Page

Legislative Internship:
* Researcher, assistant speech writer: U.S. Congressman Paul Ross

ACHIEVEMENTS:

* Community service award for three-year ongoing political series "National Economics on the Campus"

* Statistician-writer for "Citizen Alert" pamphlets on urban issues: crime, pollution, neighborhood renovation, and legal aid sponsored by Green Bay City Council

* CBS election poll assistant researcher and statistician during 1992 national elections; Regional Election Headquarters, Milwaukee, WI

CAPABILITIES:

* Work well under extreme deadline pressure
* Write fast, crisp copy
* Distill complex issues into readable, insightful prose
* Fact check with keen ability to spot inconsistencies

References and writing samples available upon request.

(1) MONICA L. SHORT
2905 Jefferson Way
Columbia, Maryland 21043

(2) General Office
Typing, Word Processing
Bookkeeping, Telephone

(3) EFFICIENT, EXPERIENCED PERSON FRIDAY ready to take full
responsibility for operating general office of small *(4)*
manufacturing, retail, or service firm. Cheerful, attractive,
and accustomed to one-person office operations. Available
because of present employer's relocation to another state.

Experience

Chesapeake Shipping Company   Baltimore, MD

Office manager and supervisor, staff of three clerks. Wrote
shipping schedule and scheduled work of 45 warehouse and
dock employees. Helped plan shipping routes. Prepared
payroll. Supervised bookkeeping, typing, telephone, and
office reception.

(5) 1987-present

Mid-States Stamping Plant   Baltimore, MD

One-person office force for this small manufacturer of auto
engine parts. Administered payroll, benefits, and record
keeping for 50 employees. Billed customers, made bank
deposits, ordered office supplies, typed all correspondence,
received callers, answered telephone, filed, and kept records
for tax returns.

(6) 1982-1987

Kramer's Furs   Washington, DC

Part-time office assistant for this 10-employee exclusive fur
retailer. Typed, filed, checked in new merchandise, kept
inventory records, assisted on selling floor when needed.

(7) 1976-1982

EDUCATION: Katharine Gibbs Executive Secretarial Course

RESUME KEY FOR MONICA L. SHORT: OFFICE MANAGER

General Issues

From the sound of her resume, Monica appears to be a middle-aged worker who is trying to get with the times. She fails by using "Person Friday," which always was a silly way to get out from the sexist "Gal Friday." With a professional title like "Office Manager," Monica will sound more nineties and less sixties. Another thing: The "cheerful, attractive" business also went the way of the "Fridays." You don't have to be dour. You just needn't say what a joyful person you are!

1. Change position of identifying block.
2. Skills not necessary in this location.
3. Omit "Person Friday."
4. Omit "cheerful, attractive."
5. Dates of employment belong at top of listing.
6. See #5.
7. See #5.

Monica L. Short
2905 Jefferson Way
Columbia, Maryland 21043
(301) 239-6271

Efficient Office Manager with 15 years' experience. Proven ability to supervise office of small manufacturing, retail, or service firm. Personable, well groomed, and accomplished as a one-person office force or directing a clerical staff.

Experience and Accomplishments

1987-present   *Clerical Supervisor*
Chesapeake Shipping Company, Baltimore, Maryland

Office manager of staff with three clerks. Schedule work of 45 warehouse and dock employees. Help plan shipping routes. Prepare payroll. Supervise bookkeeping, typing, and office reception. Evaluate and purchase word-processing software for office computers. Train employees to use it.

1982-1987   *Office Manager*
Mid-States Stamping Plant, Baltimore, Maryland

One-person office force for this small manufacturer of auto engine parts. Administered payroll, benefits, and record keeping for 50 employees. Billed customers, made bank deposits, ordered office supplies, typed all correspondence, received callers, answered telephone, filed, and maintained records for tax purposes.

1976-1982   *Part-Time Office Assistant*
Kramer's Furs, Washington, DC

Typed, filed, checked in new merchandise, kept inventory records, and assisted on selling floor when needed at this exclusive fur retail establishment.

Special Skills and Knowledge

Full-charge bookkeeping, word processing (MultiMate and WordPerfect), typing 100 wpm

Education

Katharine Gibbs Executive Secretarial Course

References

Furnished on request.

Janet Campion
11568 Canyon Park Road
San Diego, CA 92450
(619) 590-7754

(1) **OBJECTIVE:** A challenging position in the communications/marketing public relations field utilizing my demonstrated initiative, creativity, and communications abilities.

EDUCATION: University of San Diego, San Diego, California, Bachelor of Arts (2) Media Communication Studies, cum laude, May 1993. Minor: Political Science. (3) GPA: 3/64/4/00. Four years academic honors. "Who's Who Among College Students." Kappa Gamma Pi Honor Society. Department Honors.

EXTRACURRICULAR ACTIVITIES: Big Sister League, Inc. Big Sister for 10-year- (4) old girl, <u>1993 - Present.</u> College Avenue Baptist Church Volunteer. Annual usher (5) for summer Harvest Crusades. Head of toddler group two. Weekly usher for offering, <u>1991- Present.</u> Communication Studies Society. President. Responsible for organizing meetings and recruiting speakers. Led three-person team in publicity, social, and membership coordination tasks. Organize new member drives. Manage club budget, <u>1989-1993.</u> <u>VISTA</u>, campus newspaper. Staff writer. Researched and wrote weekly news articles and cover stories of current events. Attended news conference for research in Fall 1990, <u>1991-1993.</u> Volunteer (6) Resources, campus program - volunteer. Provided weekly companionship and household assistance to two elderly families, <u>1991-1992</u>. Engaged in literacy (7) building program for Hispanic youth, <u>Spring 1990</u>. Assisted maintenance of Rachel House for battered women, <u>Fall 1989</u>.

EXPERIENCE:

<u>1990 - Present</u>   **THE EASTRIDGE GROUP: ADMINISTRATIVE TEMPORARY FOR VAGABOND INNS FUNCTIONING AS MARKETING SECRETARY.**

Handle marketing, collateral, correspondence, and routing Vaga-Bond donation certificates and letters. Coordinate printing. Maintain all vendor, media, invoice, property, memo, and report files. Organize seven refurbishment parties. Attend local meetings (ConVis and Chamber of Commerce). Write press releases and assist in ad copy production, flyers, brochures, and reports. Responsible for zip code analysis, international realized group revenue, business card drawings, reports, monthly inventory. Update front desk sales guide, ticket price packages, amusement park packages.

<u>January 1993 - June 1993</u>   **KBEST OLDIES RADIO STATION: INTERN**
Promoted station at various charity events and publicized local businesses. Wrote ad copy and press releases.   (continued)

December 1992 - March 1993   **NORDSTROM: SALES CONSULTANT**
Sales for Studio 121, professional women's clothing. Participated in extensive customer service training. Follow-through on customer requests and special orders.

December 1991 - Present   **WOMEN'S HEALTH SOURCE AT SCRIPPS MEMORIAL HOSPITAL: INTERN**
Promoted free membership program for health education/guidance. Researched nationwide hospitals' memberships' programs for supervisor's article. Organized and host seminars. Responsible for OB enrollment program. Attended national health expositions to publicize and represent program to prospective members.

December 1990 - January 1991   **SAN DIEGO CONVENTION CENTER: INTERN**
Created and organized ten-volume indexed slide library of Center uses and events for publicity and mailings to prospective clients.

SKILLS: Working knowledge of Spanish. Computer literate: WordPerfect 5.0, WordPerfect for Windows, Lotus 1-2-3, Microsoft Word, Microsoft Excel, Macintosh.

⑧ **REFERENCES:**
Dr. Amanda Blakely, Professor of Communications
University of San Diego, 5998 Alcala Park
San Diego, California 92110
(619) 453-8537, extension 532

Mr. Frederick Hensley, Attorney
Brighton & Fitzpatrick
4583 Fifth Street, #200
San Diego, California 92103
(619) 884-5533

Ms. Laurie Shakeland
Vice President, Marketing & Development
VAGABOND INNS, INC.
9832 Wyandotte Avenue
La Jolla, California 93567
(619) 678-0921

RESUME KEY FOR JANET CAMPION: PUBLIC RELATIONS/ NEW GRADUATE

General Issues

Janet's resume is too long for a new graduate. An employer might look at it and think that she's been running all over campus all the time, and wonder if she has any single focus. Even though her grades are excellent and she graduated cum laude, with all of her "Extracurricular Activities" she displays too many interests. Though unusual for a recent college student, a chronological resume works for her because of her strong experience, both paid and as an intern. Adding references at the end is strictly for amateurs; Janet should know better.

1. Even though a recent college graduate, Janet can successfully replace her "Objective" with a stronger Summary.
2. Abbreviate degree.
3. Place minor next to major.
4. Omit anything referring to religion.
5. Irrelevant activity to focus of resume.
6. See #5.
7. See #5.
8. Omit references.

JANET CAMPION
11568 Canyon Park Road
San Diego, CA 92450
(619) 590-7754

SUMMARY: Recent mass media communications graduate with high honors who has consistently worked in public relations, marketing, and advertising throughout university training. Current position continues to strengthen that background. Excellent skills.

EDUCATION

University of San Diego, San Diego, CA
> BA Mass Media Communication Studies; Minor: Political Science 1993.
> CUM LAUDE; Grade Point Average: 3.64/4.00. Four-year academic honors.
> *Who's Who Among College Students*; Kappa Gamma Pi Honor Society.

ACHIEVEMENTS AND ACTIVITIES

President, Communications Studies Society: organized meetings, recruited speakers, led three-person team in publicity, social, and membership coordination. Developed new member drives. Managed organizational budget.

Staff Writer, VISTA—campus newspaper: researched and wrote cover stories, news, and features. Attended special news conference for research.

BUSINESS EXPERIENCE

1990 - Present THE EASTRIDGE GROUP: ADMINISTRATIVE TEMPORARY FOR VAGABOND INNS PERFORMING AS MARKETING SECRETARY

Write press releases. Assist in ad copy production, flyers, brochures, and reports. Represent office at Chamber of Commerce functions. Coordinate marketing, collateral, and correspondence. Handle zip code analysis, international realized group revenue, business card drawings, reports, monthly inventory update, and front desk sales guide.

1993 KBEST OLDIES RADIO STATION, San Diego, CA   INTERN
Promoted station and businesses at charity events. Wrote ad copy and press releases.

1991 WOMEN'S HEALTH SOURCE AT SCRIPPS MEMORIAL HOSPITAL,
    San Diego, CA   INTERN
Promoted free membership program for health education. Organized and hosted seminars. Coordinated OB enrollment program. Represented facility at health expos.

1990 - 1991 SAN DIEGO CONVENTION CENTER   INTERN
Created and organized 10-volume indexed slide library of center uses and events for publicity and mailings to prospective clients.

SKILLS

Computer literate: WordPerfect 5.0 and WordPerfect for Windows, Lotus 1-2-3, Microsoft Word, Microsoft Excel, Macintosh. Working knowledge: Spanish.

REFERENCES AVAILABLE UPON REQUEST

(1) Paul Kachinsky
322 Snow Meadow Lane
Burlington, VT 03261

(2) Telephone 403 217-5144

(3) PERSONAL   Married   6 ft.   190 lbs.   27 years old

(4) EDUCATION   Attended Manchester County College for two years,
majoring in Liberal Arts-Business Administration.

Attended University of Vermont, majoring in Political Science. Received Bachelor of Arts degree
in December 1982 with a 3.52 cumulative average.

Studied Stationary Engineering at Manchester
County Vocational School at night in 1984-85.
Received a Blue Seal engineer's license in August
1985.

Studied Basic Machine Shop in Manchester County
Vocational School at night in 1985-86.

SCHOLASTIC   Graduated with Honors from University of Vermont.
ACHIEVEMENTS   Was on the Dean's List for four semesters. Was
(5) one of the representatives of the Political Science Dept. to the National Model United Nations
Conference held in New York in 1982.

WORK EXPERIENCE   6/76-10/79 Produce and Frozen Foods clerk at
Johnson's Foods Plus, 435 Grayson Rd., Manchester,
VT, part-time while in school and full-time during
summers.

6/80-10/80 Canning Machine Operator at Greater
Northeastern Tank Corp. (GNTC), Henry St.,
Vosberg, VT.

(6) 4/83-7/86 Oiler at Northern Railroad, Manchester,
VT, for two years. Then promoted to Maintenance
Machinist in charge of mechanical work. Duties
also included pipefitting and operating steam
boilers, engines, lathes, and other machine shop
equipment.

Worked part-time for several years with a licensed
electrical contractor (Jones Electric Co.) installing residential and industrial services,
equipment, and wiring.

11/86-Present Maintenance Technician at Thurston
Technical Services (Data Processing Center), 531
Ensign St., Burlington, VT. Duties include Climate
Control and Building Maintenance.

BACKGROUND   (7) Brought up in Manchester area and attended local
schools. Delivered Manchester Post newspapers for
five years. Member of Manchester High School wres-
tling team for two years. Member of the Cub
Scouts and Boy Scouts for several years.

INTERESTS   (8) I enjoy reading, fresh-water fishing, camping, and
sports.

General Issues

Kachinsky is indeed an anomaly, a man with a good education who chose a blue-collar career. Nothing wrong with that—it just doesn't fit the mold. His work pattern is confusing and unclear, and the part-time and short-term stints should be removed. Better for him a functional resume that clearly lays out his achievements, followed with education that points out his academic accomplishments.

1. Center identification block.
2. See #1.
3. Omit "Personal."
4. Move and consolidate "Education."
5. Combine into "Education."
6. Experience needs to be divided along functional lines.
7. Omit.
8. Omit.

PAUL KACHINSKY
322 Snow Meadow Lane
Burlington, VT 03261
(403) 217-5144

WORK EXPERIENCE

ADMINISTRATIVE: Coordinated plant service activities, including installation, maintenance, and repair of equipment for a 30,672 ft. data processing center. Developed preventive maintenance schedules and handled all follow-through.

MECHANICAL: Responsible for repairing and maintaining all mechanical aspects of a railroad coal-dumper, including bearing replacements, pump overhauls, and general machine repairs.

PIPEFITTING: Made extensive steam line alterations and additions following a conversion from coal to #6 oil firing of three boilers totaling 1250 horsepower. Involved in replacing sections of 12-inch boiler headers.

ELECTRICAL: Assisted a licensed electrical contractor in installing residential and industrial services, equipment, and wiring.

STATIONARY ENGINEERING: Operated and maintained four piston valve steam engines; maintained four slide valve steam engines, and two duplex feedwater pumps. Also kept watch on two firetube and one watertube boilers generating 150 psi steam. Responsible for preparing this equipment for insurance inspections.

EMPLOYMENT

1986 - Present   Thurston Technical Services, Burlington, VT
Maintenance Technician

1983 - 1986   Northern Railroad, Manchester, VT
Maintenance Machinist - 1986; Oiler - 1983-1985

1981 - 1983   Jones Electrical Services, Manchester, VT
Electrician's Assistant

EDUCATION

B. A. Political Science, University of Vermont. Dean's List 4 Semesters, GPA 3.52.

Stationary Engineering, Manchester County Vocational School. Blue Seal license.

REFERENCES

Available on request.

ROBERT MANNING

(1) 23 Levering Road
Albuquerque, NM
(505) 385-1267

(2) INVESTIGATOR

(3) 37 years old, 5'10", 170 pounds, excellent health, single

(4) PRIVATE INVESTIGATOR with 17 years' experience in police, legal, insurance, and personnel investigation ; trained in law enforcement (AA degree; and skilled in reading public records and financial statements) seeks position with financial institution, law firm, detective agency, or other investigative organization. Promoted to Senior Investigator in 1991; willing to relocate to any section of United States, no dependents.

PROFESSIONAL INVESTIGATIVE EXPERIENCE:

- Investigator, New Mexico Security Patrol, Albuquerque, NM, 1983 to Present
This protective organization will phase out its investigative service at end of year and concentrate on security patrols. Invited to remain as patrol supervisor, but prefer to remain in investigative work.

During 10 years with New Mexico Security, performed following duties:

(5) Review employment records of client firms for possible risks in security contracts

Visit banks, schools, public agencies, in New Mexico, Arizona, Nevada, and California researching confidential data for client firms. (6)

Verify data presented by applicants and current information on present staffs

Follow up suspected defalcations.

- Special Agent, Market Banking Company, San Francisco, CA, 1978 - 1983
During 5 years with this international firm, was thoroughly trained in personnel background investigation, police and other public agency records, review of financial statements, interrogation, and fraud detection.

MILITARY INVESTIGATION EXPERIENCE:

Staff Sergeant, Military Police, U.S. Army, Fort Dix, NJ, and Germany 1975-1978.
Plainclothes investigation of civilian backgrounds of military personnel and sensitive problems in United States and Europe. Trained in police techniques.

(7) ## PROFESSIONAL TRAINING IN POLICE INVESTIGATION:

Associate of Arts degree in Law Enforcement, Burbank College of Government Services, Burbank, CA, June 1975

REFERENCES (8)

Full security clearance. References supplied on request.

General Issues

Manning has an interesting background and his is one of the few times that personal information is relevant. It's simply out of place at the top of the resume. Your eye wanders all over the page, trying to figure out what he does. A chronolological resume written in a linear style will make everything crystal clear.

1. Address is in wrong position.
2. Job title is in wrong position.
3. Need to reposition personal information and give it proper heading.
4. Summary has too much information of the wrong kind.
5. Poor formatting of job responsibilities.
6. No such word—means *defaults*.
7. Needs "Education" title.
8. Need separate listing for security clearance.

ROBERT MANNING
23 Levering Road
Albuquerque, NM 99551
(505) 385-1267

Private investigator: energetic and results-oriented, with 17 years' experience in police, legal, and personnel investigation. Trained in law enforcement, with academic degree. Skilled in reading public records and financial statements. Seeking full-time assignment with financial institution, law firm, detective agency, or other investigative organization.

EXPERIENCE

1983 - Present **NEW MEXICO SECURITY PATROL**, Albuquerque, NM
Senior Investigator (1991 - Present)
Investigator (1983 - 1991)

Report to general manager of the largest private security service in Albuquerque. Organization is phasing out investigative work in October. Invited to remain as patrol supervisor, but prefer to remain in investigations.
* Review employment records of client firms for security risks
* Verify applicant and current employment data for clients
* Research records in banks, school, and public agencies throughout Southwest (New Mexico, Arizona, Nevada, and California)
* Follow up suspected defaults

1978 - 1983 **MARKET BANKING COMPANY**, San Francisco, CA
Special Agent

Report to staff manager of this international investigative firm.
* Manage personnel background investigation
* Conduct public agencies' and records research
* Coordinate interrogation
* Supervise financial records review

MILITARY

1975-1978 **U.S. ARMY**, Fort Dix, NJ, and Germany
Staff sergeant, Military Police

* Investigated civilian backgrounds of military personnel and sensitive problems in United States and Europe, wearing plainclothes
* Earned security clearance
* Trained in police techniques

EDUCATION
AA, Law Enforcement, Burbank College of Government Services, Burbank, CA

PERSONAL
Single. Willing to relocate. 37 years old. 5'10". 170 lbs. Excellent health.

REFERENCES AVAILABLE UPON REQUEST

PROBLEM RESUME: PROCUREMENT/PURCHASING MANAGER/CHEMICALS

(1) RONALD SPRINGER
436 Larksdale Lane
Troy, NY 09786
(917) 673-7831

(2)
BIRTHDATE:	January 18, 1954
HEALTH:	Excellent
HEIGHT:	5' 11"
WEIGHT:	185 lbs.
MARITAL STATUS:	Married, two dependent children

OBJECTIVE:

(3) To obtain the position of Procurement/Purchasing Manager or Director with a company that promotes individual initiative and allows for individual application of management expertise.

EMPLOYMENT:

1975-Present:

Rockford Chemicals & Pharmaceuticals, Inc.

Corporate Manager - Chemicals Procurement

Responsible for managing a corporate procurement group, which purchases the major chemical raw materials for over 100 consuming plants in the United States. Commodity responsibility includes pulp and paper chemicals, plastic resins, inks, waxes, coatings, solvents, plastic film and sheet, and lignosulfonates.

1974-1975:

Rockford Chemicals & Pharmaceuticals, Inc.

Materials Manager

Designed and implemented necessary systems and procedures to establish purchasing function for B & J Chemicals, a subsidiary of National Metals & Alloys. Implemented a cost reduction program resulting in significant savings to the corporation. Coordinated purchasing activities between Corporate Purchasing and B & J Chemicals.

1969-1973:

National Metals & Alloys Company

Purchasing Agent

Was responsible for negotiating for approximately $40 million of specialty and commodity raw materials. Contributed significantly to National's cost reduction program. Performed liaison function between Corporate Purchasing and National of Canada, Ltd. Implemented program to improve reporting systems between plants and Purchasing.

General Issues

Springer's resume is a good example of giving too much of the wrong
kind of information. He needs to dig deeper to explain the significant
achievements that continue to make him *Career Progress* material. Otherwise, he can plan on staying right where he is.

1. Center identifying information.
2. Omit all this personal material. It is irrelevant today and could be
   damning tomorrow.
3. Omit and replace with "Summary."

RONALD SPRINGER
436 Larksdale Lane
Troy, NY 09786
(917) 673-7831

CAREER SUMMARY: Senior manager/chemicals responsible for procuring major chemical raw materials for more than 100 U.S. consuming plants. Designed cost reduction program currently saving over $1MM yearly. Strong team motivator and purchasing skills instructor.

EXPERIENCE

1974 - Present   **Rockford Chemicals & Pharmaceuticals**, Troy, NY
<u>Manager, Chemicals Procurement</u>

Manage a corporate procurement group, which purchases the major chemical raw materials for over 100 consuming plants in the United States. Commodity responsibilities include pulp and paper chemicals, plastic resins, inks, waxes, coatings, solvents, plastic film and sheet, and lignosulfonates. Direct six professional buyers and nonexempt employees.

Designed, developed, and implemented cost reduction programs saving over $1MM per year. Initiated program in support of Hazardous Waste Disposal project. Participated in strategy planning and negotiations for key raw materials.

<u>Materials Manager</u> (1974 - 1975)

Designed and implemented necessary systems and procedures to establish purchasing function for B & J Chemicals, subsidiary of RC&P. Coordinated purchasing activities between Corporate and B & J.

1969-1973   **National Metals & Alloys Company**, Boston, MA
<u>Purchasing Agent</u>

Performed liaison function between Corporate Purchasing and National of Canada, Ltd. Implemented program to improve reporting systems between plants and purchasing.

Negotiated for approximately $40MM of specialty and commodity raw materials. Contributed significantly to National's cost reduction program.

EDUCATION
New York State University, Rye, NY
Chemistry and Business

Also 280 hours of various management courses sponsored by Rockford Chemicals & Pharmaceuticals and National Metals and Alloys Company

REFERENCES
Available upon request.

Fred Larsen
9345 Benson Highway
Dallas, TX 75209
(214) 342-8315

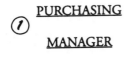

PURCHASING ①

MANAGER

③ OBJECTIVE

OFFICE
Broadway Pharmaceuticals
Marston Plaza
Dallas, Texas 75325
② (214) 521-0100

To direct all purchasing in a chemical, electronic, or other technically oriented manufacturing company.

EXPERIENCE

Twelve years' experience as purchasing manager for technically oriented manufacturers, both positions requiring a scientific academic background.

- Broadway Pharmaceuticals, Dallas, Texas, Director of Purchasing, 1990 to present. Large southern firm with 10,000 employees; being consolidated with Jackson Chemical Company, Dallas, at end of year. Invited to remain as member of purchasing team, but prefer a managerial post. Direct activities of 15 employees in Purchasing Department. Purchase $75 million annually in chemicals, production equipment, ssupplies, ④ and furniture, receiving sales people or calling upon sources. Confer with laboratory and production staffs and technical staffs of vendors. Track all sources of material and visit European, South American, and African markets.

- Hudson Electronics, Inc., Richmond, Virginia, 1979 - 1990, Advanced from Purchasing Agent to Purchasing Manager. Left this manufacturer of electronic instruments, with 3000 employees, to become Director of Purchasing at Broadway. Rapid advancement came primarily from ability to translate needs of scientific staff into high-quality, low-cost purchases.

EDUCATION
⑤ Thorough training in chemistry, electronics, and marketing.

- Master of Business Administration (Marketing), University of Virginia, Richmond, VA, July 1979.

- Bachelor of Science (Major, Chemistry; Minor, Electronics), University of Virginia, Richmond, VA, June 1975.

PROFESSIONAL AFFILIATIONS

(6) Past president of Southern United States Purchasing League (1989); Member of American Purchasing Agents' Association (1981-present); Member of Swiss American Pharmaceutical Alliance (1985-present)

(7) ## REFERENCES

Lavenia Curtis, president, Broadway Pharmaceutical Company, 456 Spruce Highway, Dallas, TX (FAX) (214) 631-0300

Henry Graves, president, Hudson Electronics, 424 Madison Avenue, Richmond, VA (FAX) (705) 862-5311

Dr. Paul Whittaker, chancellor, University of Virginia, 460 Raymond Way, Richmond, VA

General Issues

Larsen's resume looks visually interesting. But it's too tricked up. He needs to get rid of the fancy type and lines. He also needs to consolidate his experience into a format that reads faster. Because he has a very specific game plan in mind, a targeted format will work best for him. And shame on Fred—he should know better than to include such valuable and important references.

1. Omit strange positioning of title.
2. Because employer knows of his plans to leave, it is permissible to include his office number. It should be positioned differently, on same line as home address.
3. Omit objective; use summary.
4. Misspelling.
5. Omit this line; his degrees spell this out.
6. Put affiliations in linear format for ease of reading.
7. Omit references.

RESUME SOLUTION: TARGETED, PURCHASING MANAGER

FRED LARSEN
9345 Benson Highway
Dallas, TX 75209
(214) 342-8315

Office:
Broadway Pharmaceuticals
Dallas, TX 75325
(214) 521-0100

JOB TARGET: Purchasing Manager in chemical, electronic, or other technically oriented manufacturing facility, for current manager with strong scientific educational background and 12 years' professional experience.

CAPABILITIES:

* Translate complex needs from scientific team into practical requirements for this large pharmaceutical manufacturer with 10,000 employees.
* Coordinate purchasing functions of each agent into cost-effective whole.
* Oversee all RFQ's.
* Research sources worldwide, with excellent suppliers developed in Europe, South America, and Africa.
* Confer with laboratory and production staffs and technical staffs of vendors.
* Develop technical specifications.

ACHIEVEMENTS:

* Direct activities of 15 employees in purchasing department.
* Purchase $75 million annually in chemicals, production equpment, supplies, and furniture.
* Reduced purchasing costs on an average of 20% during past three years.

EMPLOYMENT:

1990 - Present, Broadway Pharmaceuticals, Dallas, TX   Purchasing Manager
1979 - 1990, Hudson Electronics, Richmond, VA   Purchasing Agent & Purchasing Mgr.

PROFESSIONAL AFFILIATIONS:

* Southern United States Purchasing League - Past President, 1989
* American Purchasing Agents Association - Member - 1981 to present
* Swiss American Pharmaceutical Alliance - Member - 1985 to present

EDUCATION:

* MBA (Marketing), University of Virginia, Richmond, VA
* BS (Chemistry major, Electronics minor) University of Virginia, Richmond, VA

REFERENCES:

Excellent references supplied upon request. Because current employer is being consolidated, and I have chosen not to remain as a member of the purchasing team, my employer knows of my decision to leave. Therefore, you may contact me during business hours.

DAVID GOLDMAN
16342 Carmel Highlands Road, #150
Hollister, CA 95367
①

1993-1994:

Total Quality Management Consultant: Performed Baldrige assessment of IBM Software Manufacturing, Boulder, CO, preparatory to their formal application for the IBM Silver Quality Award. IBM Boulder subsequently received the Silver Award.

Administrator/Business Manager: Responsible for planning, setting up, and implementing all financial and business activities for law firm in Hollister.

1991-1993:

Program Manager, Total Quality Management, IBM Santa Teresa Laboratory: Responsible for planning and implementing annual comprehensive site-wide TQM assessments and evaluations. Using representatives from all site areas, the Malcolm Baldrige National Quality Award Criteria, and local innovative approaches, the TQM goals were achieved each year. In 1991, the Lab was awarded the IBM Silver Quality Award. In 1992, the Lab achieved the maximum Variable Pay measurement (the 1992 measurement was based on TQM). In 1993, the Lab received the rarely awarded IBM Gold Quality Award.

Baldrige Consultant, IBM Santa Teresa Laboratory: Served as the Lab's primary expert on the Malcolm Baldrige Award Criteria. Responsibilities included teaching, conducting workshops, consultation with all of the Lab's Development areas, and consultation to other IBM locations.

1990:

Site Manager, IBM Menlo Park: Responsible for managing the 300-employee software development location during location transition from Menlo Park to the Santa Teresa Laboratory located in south San Jose. In addition to the customary operational responsibilities, a significant task was to find optimum placements for the Menlo Park population. This was accomplished with an attrition rate below nonrelocating sites.

1981-1989:

Operations Manager, IBM Menlo Park: Responsible for the day-to-day operational management of the 5-building, 300+-employee software development laboratory in Menlo Park, CA. Specific responsibilities included: Building/Facility management, Information Systems (I/S), Site-wide Planning, Technical Editors, Security, Safety, Administration, and Space.

1975 - 1980:

Center Planning Manager, IBM Menlo Park: Responsible for gathering the input and preparing the Site's Strategic and Operational Plans. Tasks included business forecasting, resource planning, and business case evaluations.

1970-1975:

Health Industry Planning Manager, IBM Systems Development Division (Kingston, NY) and Health Industry Planning Manager, IBM Advanced Systems Development Division (Yorktown Heights, NY): Responsible for the worldwide strategy and planning of the IBM Health/Medical products under development in the Systems Development and the Advanced Systems Development Divisions. In 1972, served 6 months as Administrative Assistant to the General Manger of the Advanced Systems Development Division. From 1973-1975 was additionally responsible for the Division's Retail Rubber Industry Planning.

1966-1970:

Development Manager, Medical Information Systems, IBM Advanced Systems Development Division (Yorktown Heights, NY) and Development Manager, Medical Development, IBM Data Processing Division (White Plains, NY): Responsible for the development of medical applications including Shared Hospital Accounting System for the IBM 360 (SHAS) and applications for the Hospital Clinical Laboratory, Dietary Service, Admitting Room, and Medical Records. Also, led a large, comprehensive joint study with the Roosevelt Hospital in New York City.

1964-1966:

Advisory Systems Engineer, Medical Applications, IBM Data Processing Division (Los Angeles, CA): Responsibilities included heading a leading-edge applications joint study with the Long Beach Memorial Hospital.

1961-1964:

Systems Engineer Los Angeles Data Center, IBM Data Processing Division (Los Angeles, CA): Responsible for technical leadership of the IBM 7074 Computer System and also served as the technical assistant to the Data Center Manager.

1960-1961:

Applied Science Representative, Los Angeles Wilshire Branch Office, IBM Data Processing Division (Los Angeles, CA): Provided technical assistance to customers, programmed a matrix inversion routine for Mobil Oil, and taught FORTRAN to Chevron.

1958-1960: ②

Instructor, Dept. of Mathematics, University of Southern California (Los Angeles, CA): Taught Calculus and Business Mathematics to undergraduates while taking graduate courses aimed at Ph.D. in Mathematics.

EDUCATION: ③

B.S. in Mathematics, University of Southern California
Completed 32 graduate mathematics units towards Ph.D.

General Issues

Goldman's is the classic case of a technical expert who spent his entire career with a single firm. His working materials and background are excellent, his presentation too wordy. Many paragraphs and sentences are too long. Format does not move fast enough. Many years need to be lopped off his resume. Now considering a shift into consultation following IBM retirement, Goldman's resume can follow either a chronological or targeted format.

1. No telephone number.
2. Abbreviate name of this well-known university.
3. Incorrect verbiage in this line about his degree.

DAVID GOLDMAN
16342 Carmel Highlands Road #150
Hollister, CA 95367
(408) 745-3390

SUMMARY: IBM Total Quality Management (TQM) program manager and facilities manager for 13 years, plus extensive development experience in IBM's medical applications systems development. Managed sites with up to 300 employees. Known as efficient facilitator and team player with broad perspective who cuts through bureaucracy to achieve desired results for corporate facilities. Advanced mathematics background.

EMPLOYMENT EXPERIENCE:

<u>1993 - Present</u> <u>Consultant/Business Planner,</u> Hollister, CA
Consulted with IBM Software Manufacturing, Boulder, CO, on TQM. Performed Baldrige Assessment that led to facility's winning the IBM Silver Quality Award. Directed financial and business planning for one of Hollister's largest law firms.

<u>1991-1993</u> **IBM, Santa Teresa, CA**
<u>Program Manager, TQM, and Baldrige Consultant:</u> Planned and implemented annual comprehensive site-wide TQM assessments and evaluations. Met goals every year. In 1993, Lab received rarely awarded IBM Gold Quality Award. In 1992, Lab achieved maximum possible pay increases, based solely on TQM. In 1991, Lab won IBM Silver Quality Award. As Baldrige consultant, conducted workshops, taught, and consulted with all Lab Development areas. Consulted with other IBM locations.

<u>1975 - 1990</u> **IBM, Menlo Park, CA**
<u>Site Manager (1990):</u> Managed 300-employee software development center during transition from Menlo Park to Santa Teresa. Directed optimum placements for Menlo Park employees, resulting in an attrition rate below nonrelocating sites.

<u>Operations Manager (1981-1989):</u> Directed day-to-day operations at - building 300+-employee software development facilities including building/facility management, information systems (I/S), site-wide planning, technical editors, security, safety, administration, and space.

<u>Center Planning Manager (1975-1980):</u> Prepared site's strategic and operational plans including business forecasting, resource planning, and business case evaluations.

<u>1970- 1975</u> **IBM Systems Development Division, Kingston, NY, and**
IBM Advanced Systems Development Divison, Yorktown Heights, NY
<u>Health Industry Planning Manager:</u> Responsible for worldwide strategy and planning of IBM Health/Medical products under development in these two centers.

P. 2. Resume - David Goldman

<u>1966-1970</u> **IBM Advanced Systems Development Divison, Yorktown Heights, NY,
and IBM Data Processing Division (White Plains, NY)**
<u>Development Manager:</u> Headed development team for medical applications including
Shared Hospital Accounting System for the IBM 360 (SHAS) and applications for Hospital
Clinical Laboratory, Dietary Service, Admitting Room, and Medical Records. Led comprehensive joint study with Roosevelt Hospital, NYC.

EDUCATION:

Completed 32 graduate mathematics units toward Ph.D.

Taught Calculus and Business Math to undergraduates while taking graduate courses

B.S. Mathematics, USC, Los Angeles, CA

REFERENCES SUPPLIED UPON REQUEST

DAVID GOLDMAN
16342 Carmel Highlands Road #150
Hollister, CA 95367
(408) 745-3390

JOB TARGET: Consulting positions with major corporations in which I can use my experience at IBM in Total Quality Management and Baldrige specialization.

CAPABILITIES:

* Take-charge specialist recognized company-wide for achieving results and cutting through bureacracy.
* Hands-on management skills.
* Team player; group facilitator.
* Teacher and workshop conductor.

ACHIEVEMENTS:

* Consulted and directed Baldrige assessment of IBM Software Manufacturing, Boulder, CO, that led to receipt of IBM Silver Quality Award in 1993.
* Planned and implemented annual comprehensive site-wide TQM assessments and evaluations at IBM Santa Teresa, CA, with the following results:
  -1993- Won rarely awarded IBM Gold Quality Award.
  -1992- Achieved maximum Variable Pay measurement for facility, with measurement based on TQM.
  -1991- Earned IBM Silver Quality Award.
* Conducted workshops, taught, and consulted with all Lab development areas, and consulted with other IBM locations.

PROFESSIONAL EMPLOYMENT:

1991- 1993 IBM, Santa Teresa, CA   Program Manager and Baldrige Consultant
1975 - 1990 IBM, Menlo Park, CA   Site Manager, Operations Manager, Center
                                   Planning Manager
1970 - 1975 IBM, Systems Development Division, Kingston, NY, and
            Advanced Systems Development Division, Yorktown, NY
                                   Health Industry Planning Manager
Plus additional positions in management and systems engineering at IBM

EDUCATION:

* Completed 32 graduate mathematics units toward Ph.D.
* Taught Calculus and Business Math to undergraduates while taking graduate courses
* B.S. Mathematics, USC, Los Angeles, CA

REFERENCES SUPPLIED UPON REQUEST

(1)

Sylvan Harcourt
624 W. Washington Blvd., #43
Los Angeles, CA 90007
(213) 873-5889

(2) OBJECTIVE:

To obtain a management position commensurate with my abilities and my experience in a growing and dynamic organization with opportunity for advancement by merit.

(4)

(3) WORK HISTORY:

University of Southern California   8/90 - Present
OPERATIONS MANAGER - DINING SUPPORT SERVICES

(5) * Managed full-service restaurant campus saloon, which averages over 500 lunches daily. (6)

* *Designed, implemented, and administrated an all-campus, prepaid debit card used primarily for dining - realizing over 30% annual growth during my administration with 1993 deposits totaling over $15,000,000.*

(7) * Served as project manager for creation of on-campus bar and saloon.

(8) * Responsibilities: supervision of operations, inventory control, staffing and training, budgetary control, and creation of formal business and marketing plans.

(9)

Service America - Cal State Fullerton   8/88-8/90
FOOD SERVICE MANAGER

* Managed high-volume campus pub and sidewalk cafe.

* *Initiated promotional campaigns and concert functions that resulted in 300% revenue growth in first year.*

(10) * Responsibilities: cost/cash control, inventory, training, marketing, and promotions.

(11)

32nd Street Cafe and Saloon   1/83 - 8/88
ASSISTANT RESTAURANT MANAGER/BEVERAGE MANAGER

* Assistant Manager and Bar Manager for fast-paced Los Angeles restaurant/bar with revenues over $2,000,000 annually.

* *Orchestrated and executed plans to accommodate and capitalize upon dramatic increase in business during the 1984 Olympics.*

(12) * Responsibilities: supervision of restaurant and bar operations, direct accountability/responsibility for all decisions pertinent to bar operations.

(13)

Restaurant Adventures, Inc. (varied locations)   5/77 - 1/83
BEVERAGE MANAGER/TRAINING TEAM MANAGER

* Beverage Manager for popular Southern California restaurant group (Quiet Cannon, Orange County Mining Co., and others)

(14) * In addition to standard operational responsibilities, acted as "New Unit Opening Manager," effectively recruited and trained staff, implemented policies for 4 new units.

(15)

EDUCATION:

University of Southern California
B.S. Business Administration - Marketing Emphasis (degree pending)
3.23 GPA - while working full time

References available upon request

RESUME KEY FOR SYLVAN HARCOURT: RESTAURANT MANAGER

General Issues

Resume is basically a good one that needs refining. Harcourt has chosen the correct format to exploit his considerable achievements. Though a student who is just about to graduate, Harcourt's experience is so strong that it made the difference in placing his work background *before* his educational accomplishments.

1. Name should be capitalized.
2. "Objective" becomes a tightly focused summary of his achievements.
3. "Work History" is incorrect phraseology: "Experience" is sufficient.
4. Incorrect positioning of employment dates weakens resume.
5. Correct phrasing in this linear resume includes information on Harcourt's superiors and background about the nature of the business.
6. No such word as "administrated." Should be "administered."
7. Covered in job title; positioning here is redundant.
8. Belongs in opening subparagraph; see #5.
9. See #4.
10. See #5.
11. See #4.
12. See #5.
13. See #4.
14. See #5.
15. Layout of education needs reformatting.

SYLVAN HARCOURT
624 W. Washington Blvd., #43
Los Angeles, CA 90007
(213) 873-5889

Innovative restaurant/bar manager with over 15 years' experience directing lively youth-oriented facilities. Successful, creative marketing planner. Self-motivated hard worker.

EXPERIENCE:

1990- **UNIVERSITY OF SOUTHERN CALIFORNIA, LOS ANGELES, CA**
Present **Operations Manager—Dining Support Services**
Report to General Manager of this full-service restaurant/campus saloon averaging over 500 meals daily. Responsibilities: operations supervision, inventory control, staffing, training, budget controls, business and marketing plans.
* Designed, implemented, and now administer all-campus, prepaid debit dining card, with over 30% annual growth and 1993 deposits exceeding $15,000,000.

1988- **SERVICE AMERICA—CAL STATE FULLERTON, FULLERTON, CA**
1990 **Food Service Manager**
Reported to Supervisor, dining services for this high-volume campus pub and sidewalk cafe. Responsibilities: cost/cash control, inventory, training, marketing.
* Initiated promotions and concerts resulting in 300% revenue growth in first year.

1983- **32ND STREET CAFE AND SALOON, LOS ANGELES, CA**
1988 **Assistant Restaurant Manager/Beverage Manager**
Reported to manager for fast-paced Los Angeles restaurant near USC campus grossing over $2,000,000 annually. Responsibilities: supervision of restaurant and bar operations, direct accountability/responsibility for all bar operations.
* Orchestrated and executed plans to capitalize upon dramatic volume increase due to proximity to L.A. Coliseum during 1984 Olympics.
* Business volume rose by 35% during the 3 months before, during, and after games.

1977- **RESTAURANT ADVENTURES, THROUGHOUT SOUTHERN CALIFORNIA**
1983 **Beverage Manager**
Reported to Bar Manager for popular Southern California restaurant group that included the Quiet Cannon, Orange County Mining Co., and others.
Responsibilities: operations, plus "New Unit Opening Manager."

EDUCATION:

U.S.C. Business Administration—Marketing Emphasis (degree expected 1994)
3.23 GPA—while working fulltime

REFERENCES AVAILABLE UPON REQUEST

PROBLEM RESUME: SALES/FINANCIAL/NEW GRADUATE

(1)

Victoria Ramos
584 W. 52nd Place
Los Angeles, CA 90043
(213) 298-5783

Professional Objective
To pursue a career in international business as a financial sales or promotional representative.

Education
Bachelor of Arts, English Literature

University of California, Los Angeles, Summer 1992

Course work includes: Microeconomics, Macroceconomics, Statistics, Differential Calculus, Intermediate Financial Accounting, and Computer Information

Systems knowledge of BASIC, Lotus 1-2-3, Latex, and various Macintosh programs including Microsoft Word 5.0, Excel 2.2, and MacDraw. Passed the CBEST Examination, 1991

(2) Attended the Wordsworth Summer Conference 1992: an extensive two-week study of the Romantic Literature period in Grasmere, England. Later traveled in France and Italy.

(3) Foreign languages spoken: conversational French and Tagalog (Philippine lang.)

Experience
Admin./Accounting Asst., Electrical Engineering Dept., UCLA Oct. 1991 - present

Assistant to the Director of the Center for High Frequency. Perform accounting procedures to process check requests, travel advances, reimbursements, travel vouchers, purchase orders, and expenditure adjustments. Verify ledger expenditures, reconcile variances, and prepare reports on expenditures. Prepare scientific manuscripts for publication, which required typesetting, editing, and proofreading. Knowledgeable in fund accounting, and have some exposure to contracts and grants.

Payroll Assistant, Electrical Engineering Department, UCLA Nov 1989 - Sept 1991

Compiled timesheets for the bi-weekly and monthly payroll; managed records of vacation and sick time accruals; and updated database files.

Telemarketing Agent, Annual Alumni Fund Campaign UCLA Oct. - Dec 1989

Bank Teller, Security Pacific National Bank, #247, Torrance, CA May - Sept. 1989

Customer Service Asst., R.R. Donnelly & Sons, Inc., Torrance, CA June - Aug 1989

Assistant to the Supervisor of the Shipping Department and Customer Service representative of this international publishing company. Duties included answering customer inquiries, word processing, and updating database files on the IBM computer using Lotus 1-2-3.

General Acct. Ashby Jewelers Inc., Del Amo Branch, Torrance, CA Jan.- Oct, 1989

Processed billing, account receivables, and account payables; analyzed customer credit eligibility; prepared daily balances, weekly balance sheets, and end-of-month records.

Honors and Awards
UCLA College of Letters and Science Honors Div., 1989; UCLA Student
(4) Research Prog. Stipend Award Recipient - Winter, Spring 1989
Commencement Speaker for the Graduating Class of 1992, Humanities Division University of California, Los Angeles

Activities **(5)** UCLA Speech and Debate Team, 1989; UC Dance Theater, 1991 and 1992; Member, L.A. Classical Ballet; attended Royal Winnipeg Professional Division School, Sum. 1991

General Issues

Ramos has very impressive educational credentials, but they are split into two sections and hidden through too-small typeface. Her work experience lends credence to her stated career path, but she has too many jobs in too short a period (even given the fact that they were summer positions). Her resume layout makes you look too hard to associate duties with job titles. Since she *is* a recent graduate, having a functional resume with job duties categorized would be a much better plan.

1. Name should be capitalized.
2. Section should be moved lower down. Travel is relevant because of her international career goals.
3. Section should be moved lower down.
4. Impressive credentials belong near top of resume.
5. Dance background is irrelevant.

VICTORIA RAMOS
584 W. 52nd Place
Los Angeles, CA 90043
(213) 298-5783

SUMMARY: Recent UCLA graduate with strong educational and work-related economics and financial background. Multilingual, with foreign travel experience for international financial sales/promotional position.

EDUCATION

1992, BA, English Literature, UCLA,
    Commencement Speaker, Humanities Division
1989, UCLA College of Letters and Science Honors Divison
    Student Research Program
    Stipend Award Recipient
    Speech and Debate Team Member
    Business/Financial Coursework: Microeconomics, Macroeconomics, Statistics, Differential Calculus, Intermediate Financial Accounting

EXPERIENCE

ACCOUNTING AND HUMAN RESOURCES Process check requests, travel advances, reimbursements, travel vouchers, purchase orders, and expenditure adjustments. Verify ledger expenditures and prepare reports; reconcile variances. Knowledgable: fund accounting, contracts, and grants. Manage retail billings, accounts receivable and payables. Prepare daily and weekly balances and end-of-month records. Compile payroll timesheets. Manage vacation and sick time accruals with relevant database updates.

TECHNICAL KNOWLEDGE AND SKILLS

Computer information systems knowledge includes: BASIC, Lotus 1-2-3, Latex, and various Macintosh programs including Microsoft Word 5.0, Excel 2.2, and MacDraw. Passed CBEST Examination, 1991.

LANGUAGES AND TRAVEL

Speak: Conversational French. Tagalog (Philippine national language).
Travel: England for Wordsworth Summer Conference. Canada, France, Italy.

PRE- AND POSTGRADUATION EMPLOYMENT

1991 - Present   UCLA Electrical Engineering Dept., Los Angeles, CA
        Administrative/Accounting Assistant
            Payroll Assistant (1989 - 1991)
1989   Security Pacific National Bank, Torrance, CA, Teller
1989   Ashby Jewelers, Inc., Torrance, CA, General Accountant

(1) MILTON WELLS

   346 Lakeview Road   (2) Five Years...
   Indianapolis, IN 43908
   (315) 478-0367   Sales Success

(3) <u>JOB OBJECTIVE:</u>   To sell sophisticated products to
                 profesional buyers - physicians, scientists,
                 and educators

(4) <u>RECORD OF ENTERPRISING SALES ACHIEVEMENT</u>

   <u>Pharmaceutical Sales Representative</u>, Collins Corp.,
   Chicago, IL   1990-present.

      Call upon physicians in Indiana territory, averaging
(5)   eight to ten calls a day, displaying and explaining
      new products and answering all questions, or getting
      answers to these questions from the home office.

      Keep myself knowledgeable on Collins' 1000-product
      line

(6) Desire change because career advancement opportunities
   are limited in this company.

   <u>College Textbook Sales Representative</u>, Walston-Green
   Publishing Co., Boston, MA   1986-1990.

      Displayed catalog of titles and promotional materials
(7)   to department heads and professors in all colleges
      throughout Massachusetts, answering questions and
      urging adoptions.

      Solicited manuscripts for new textbooks.

(8) During my four years with Walston-Green, sales in
   Massachusetts increased 25% each year. I left because
   Collins offered a high salary with opportunity to advance
   to sales manager.

<u>EDUCATION</u>
   <u>Bachelor of Arts</u>, Liberal Arts, Princeton University,
   1986.
   Strong background in humanities with all electives in
   chemistry.

<u>REFERENCES</u>
   Full references will be furnished on request.

General Issues

Maybe it's a salesman's natural bent to pat himself on the back to keep his morale up, but Wells has a tone in this resume that will keep head-hunters and employment managers far from his door. He editorializes, when he should just let the facts speak for themselves. Before rewriting, Wells needs to go back and review his achievements at his current position.

He also displays poor judgment in commenting on his reasons for desiring change. He left his first stated position because he had an opportunity to become sales manager. Nearly five years into that current position he still hasn't advanced and now says "opportunities are limited." An employer might wonder if Wells really has what it takes to advance.

1. Incorrect positioning of identifying information.
2. Gratuitous editorializing is totally out of place.
3. Omit objective.
4. See #2.
5. Tone is too laid back.
6. Omit reason for leaving.
7. See #5.
8. See #6.

MILTON WELLS
346 Lakeview Road
Indianapolis, IN 43908
(315) 478-0367

Sales professional with an eight-year track record of selling pharmaceuticals and college-level textbooks to sophisticated buyers. Consistently meet or exceed monthly quotas. Tenacious dedication to making the sale. Self-motivated and well-organized individual, ready to move up to management.

EMPLOYMENT EXPERIENCE

1990 - Present   <u>Pharmaceutical Sales Representative</u>
                    COLLINS CORPORATION, Chicago, IL

* Call on eight to ten physicians' offices daily throughout state of Indiana
* Provide clear, concise oral information to physicians' staffs and MDs
* Introduce new products on average of two per month
* Research medical questions for MDs through home office
* Study Collins' catalog and updates to stay abreast of 1000-product line

   <u>Achievements:</u>

   * Received Sales of the Month Award in two consecutive years
   * Made or exceeded quotas every month last year

1986 - 1990   <u>College Textbook Sales Representative</u>
                    WALSTON GREEN PUBLISHING CO., Boston, MA

* Met with department heads and university professors throughout Massachusetts
* Displayed catalog of titles and promotional materials with emphasis on new offerings
* Solicited manuscripts for new textbooks

   <u>Achievements:</u>

   * Increased sales by 25% throughout Massachusetts each year of my tenure

EDUCATION

Princeton University, BA, Liberal Arts
Strong background in humanities; all electives in chemistry.

REFERENCES

Full references will be furnished on request.

PROBLEM RESUME: SECRETARY

(1) WANDA PELSTON
    9743 Evans Road
    Long Island, NY 05732
    (614) 888-5430

(2) EXECUTIVE SECRETARY

(3) OBJECTIVE: To serve as executive secretary to first- or second-echelon officer of major corporation.

SUMMARY: Twenty years' experience as secretary, fourteen of these with top corporate executives. Poised, resourceful, excellent letter and report writer. Excellent recommendations and references.

EXPERIENCE: Executive secretary and administrative assistant, (4) executive office, XL Atlantic Building, Inc. NYC.

1987-Present   This large manufacturer of building materials with (5) 15,000 employees has been purchased by SMI Plastics; executive office being phased out. Hired as (6) secretary to president; placed in charge of executive-office clerical staff (15 employees) when president was also given assignment of board chairman.

    Serve as administration assistant to president, arrange all board and executive staff meetings, preparing agendas and covering minutes; assist in preparation of all major reports and directives issued by executive office; prepare routine correspondence using WordPerfect 6.0 for Windows for signature of president; supervise clerical staff.

1977-1987   Executive secretary to president, Jones Steel Corp., (7) Pittsburgh, PA. Assisted the president of this large corporation (30,000 employees) in his administrative duties; arranged meetings and trips; wrote for his signature all routine letters and reports; and maintained his office in his absence.

1974-1977   (8) Secretary to sales staff. Royston Fabrics, Pittsburgh, PA. Served staff of five textile salesmen, writing correspondence and typing records.

EDUCATION: (9) Attended basic and advanced executive secretarial one-week seminars of American Management Assn., New York City, 1985 and 1987.

    Graduate, Lyons High School, Pittsburgh, PA 1974 Valedictorian of class.

REFERENCES   References will be forwarded on request.

General Issues

Pelston's resume lacks the crisp authority you'd expect of an efficient personal secretary. Layout has no clear, delineating text to show her powerful position. Her language is also fuzzy. She omits technical skills. Even though her high-level responsibilities are more administrative than functional, skills are expected on a secretarial resume.

1. Incorrect positioning of information section.
2. Incorrect positioning of title.
3. Omit objective.
4. No clear, identifiying format for employer's name, and her position.
5. This is one of the few times that putting a reason for leaving is acceptable. The firm is so large that news of its purchase will have made the major business trades. Her reference to this change at the top, therefore, will be understood by prospective employers.
6. Too wordy.
7. See #4.
8. See #4.
9. Incorrect format for educational information.

WANDA PELSTON
9743 Evans Road
Long Island, NY 05732
(614) 888-5430

CAREER SUMMARY: Twenty years of continuous professional secretarial work, 17 of them with top corporate executives of major national corporations. Poised, resourceful, excellent letter and report writer. Extensive clerical staff supervisory responsibilities.

EXPERIENCE

1987-Present **XL ATLANTIC BUILDING, INC.** New York, NY
<u>Executive Secretary and Administrative Assistant</u>
Personal secretary to and report to president of this large manufacturer of building materials with 15,000 employees. Placed in charge of executive-office staff of 15 employees when president also received assignment as Chairman of the Board. (Purchase by SMI Plastics will lead to realignment and displacement of executive office staff.)

* Arrange all board and executive staff meetings including agenda preparation.
* Attend meetings to take minutes.
* Assist in preparation of all major reports and directives from executive offices.
* Prepare routine correspondence for signature of president.
* Supervise clerical staff.

1977-1987 **JONES STEEL CORPORATION**, Pittsburgh, PA
<u>Executive Secretary</u>
Personal secretary to and reported to president of this major steel corporation with 30,000 employees.

* Arranged president's daily schedule for meetings and trips.
* Wrote president's routine correspondence for his signature.

1974-1977 **ROYSTON FABRICS**, Pittsburgh, PA
<u>Secretary to sales staff</u>

PROFESSIONAL SKILLS

Computer: <u>Hardware</u> IBM PC; WordPerfect 5.0 and WordPerfect 6.0 for Windows. Keyboarding 90 wpm. Shorthand 120 wpm. Dictaphone proficiency.

EDUCATION

Basic and advanced secretarial seminars, American Management Assn., NY, 1985 and 1987.

REFERENCES FORWARDED UPON REQUEST

PROBLEM RESUME: SOCIAL WORKER

(1) PROFESSIONAL EXPERIENCE AND TRAINING OF CELIA SWANSON, SOCIAL WORKER

(2) 27 First Avenue, White Plains, New York 21032 (914) 883-5672

(3) Social Worker with RN seeks position with urban organization that needs combination of social work and nursing insights.

Experience in Social Work

1991-Present **Family Caseworker,** Nondenominational Charities, Inc., White Plains, NY. Interviewed members of families at
(4) Nondenominational Hospital or at their homes, investigated needs, proposed plans for assistance, followed up on assistance offered.

1983-1991 **Medical Caseworker,** Trenton Department of Hospitals, Trenton, NJ. Interviewed families of patients to
(5) determine financial competence, arranged structure of fees to be paid by family, arranged for outpatient follow-up, counseled families for outpatient and convalescent cooperation.

Education

1980-1983 **Bachelor of Science, Social Work, Manhattan University School of Social Work**. Concentrated training in all
(6) areas of social work with varied short internships in pediatric, psychiatric, and rehabilitation areas.

1977-1980 **Registered Nurse**, Trenton Hospital, Trenton, NJ
(7) Full training in all phases of general nursing with special course in psychiatric nursing. Realized from experience that true interest lies in social work.

Professional Affiliations

National Association of Social Workers and Medical Caseworkers Society.

Interests

(8) **Languages**. Have conversational and reading command of Spanish and Italian. Good reading knowledge of French and German, with fair conversational ability. Travel to Puerto Rico and Mexico often to improve and maintain Spanish.

References

(9) References covering all phases of education and experience available on request.

RESUME KEY FOR CELIA SWANSON: SOCIAL WORKER

General Issues

Outside of some excess verbiage at the top, nothing is terribly wrong with Swanson's resume. It just lacks the spark that a linear format would give by highlighting important elements of her career.

1. Omit excess verbiage.
2. Put address and phone into correct format.
3. Omit objective; use summary.
4. Edit and tighten.
5. See #4.
6. See #4.
7. See #4.
8. Too wordy; see #4.
9. See #8.

CELIA SWANSON
27 First Avenue
White Plains, NY 21032
(914) 883-5672

SUMMARY: Urban casework specialist with 11 years continuous experience. Strong interviewing skills. Adept at translating on-the-scene rapport into support paperwork that generates desired placements and assistance. Emotionally mature with special insights obtained through RN background.

EXPERIENCE:

1991-Present   <u>Family Caseworker</u>, NONDENOMINATIONAL CHARITIES, INC.,
White Plains, NY
Report to Social Services Manager of one of the largest family assistance programs in White Plains, linked to hospital intake.

* Interview family members at Nondenominational Hospital or at homes.
* Investigate needs and propose plans for assistance.
* Counsel and guide females, with special involvement in battered women and children.
* Provide regular follow-up and reports.

1983-1991   <u>Medical Caseworker</u>, TRENTON DEPARTMENT OF HOSPITALS,
Trenton, NJ
Reported to Senior Manager.

* Interviewed patients' families to determine financial competence.
* Arranged structure of fees to be paid by families.
* Planned outpatient follow-up and referred to local services.
* Counseled families for outpatient and convalescent cooperation.

EDUCATION:

B.S. Social Work, Manhattan University School of Social Work, New York City
Internships in pediatric, psychiatric, and rehabilitation areas.
RN, Trenton Hospital, Trenton, NJ
Full training: general nursing. Specialization in psychiatric nursing led to social work career path.

ORGANIZATIONS:

National Association of Social Workers and Medical Caseworkers Society.

LANGUAGES:

Conversational: Good: Spanish, Italian; Adequate: French, German.
Reading: Good: French and German.

REFERENCES AVAILABLE UPON REQUEST

(1) <u>Daniel J. O'Keefe</u>
MIS professional with strong mainframe background

(2) 25 Sunnyside Lane
Walnut Creek, California 94596   (415) 965-5418

Objective: To serve as senior systems analyst in computer
                  center of a large corporation, where my knowledge of
                  sophisticated systems, procedures, and database
                  administration can be applied to challenging
                  problems.

<u>EXPERIENCE:</u>   Five years of high-level **(3)** systems design experience
                  with THE BERKELEY GROUP, Oakland, California, a
                  large management consulting firm, as Manager,
                  Systems Design, 1985-present.

*Consulted with clients, reviewed problems, and developed proposals,
specifying adaptive uses of present hardware or new hardware to
meet systems needs. Designed new systems, oversaw installation and
initial operation, including orientation and training of client
personnel.

*Developed and implemented systems designs and programs for client
companies, which involved close contact with senior management of
client firms to develop objectives, review constraints, and
recommend appropriate systems development. Built mathematical models
and programs.

(4) REASON FOR LEAVING: To find a new position with a larger
corporation.

                  Five years of solid programming and systems design
                  experience with ELECTRONIC SOLUTIONS CORP., Irvine,
                  California, a small but important software
                  manufacturer, 1980-1985. **(5)**

*Led the team developing Write-Right, a new word-processing program
with grammar and spelling check features that has met with great
market success.

*Redesigned ESC's order entry and shipping program, which resulted
in a 50% reduction in order-to-ship time.

<u>EDUCATION</u>   Master of Science (M.S.) Computer Sciences,
                  University of California at Irvine, 1981. Thesis:
          (6) "Computer Programming: A Systems Approach" published
                  by University Press.
                  Bachelor of Science, Electrical Engineering,
                  California Polytechnic Institute, 1979. Full training
                  in electrical engineering with heavy minor in

mathematics. Learned mathematical model building in a trailblazing program that anticipated total systems approach to computer design.

<u>Graduated summa cum laude, Member Phi Beta Kappa</u>

AFFILIATIONS   Member of three national professional societies:

*National Society for Systems Analysts (NSSA)
*American Association of Electric Engineers (AAEE)
*American Association of Computer Scientists (AACS)

(7) INTERESTS   Electronic Music, Programming Computer Games, Mountain Climbing, Philosophy, especially Epistemology

(8) PERSONAL   Born in Alameda, California; attended California public schools, except for one year in Spain as an exchange student; speak and write fluent Spanish; single; no children; two dogs and a cat.

(9) REFERENCES   Too numerous to list. A list of professional and academic references will be provided upon request.

General Issues

O'Keefe has impressive credentials, but they're buried under an avalanche of too many words and too much unnecessary, inappropriate information. He highlights experience and fails to shine any light on accomplishments that make him stand out. He editorializes with phrases like "strong" background, "high-level" experience, and the "large" size of previous employers. He should use his mathematical ability to quantify his achievements and make the reader understand his performance level. A crowded page layout where key dates and information gray out also works against him. He needs to review his quantifiable successes and rewrite in a chronological, linear format where the information will hit the reader immediately.

1. Inappropriate separation of name from balance of identifying block
2. See #1.
3. Editorialization. Needs to be more specific.
4. Omit reason for leaving.
5. See #3.
6. Education is a jumble and needs to be clarified.
7. Omit interests. It may label him a kooky California scientist.
8. Extraneous, unnecessary information.
9. Gratuitous comment. "Too many to list" suggests the whole world knows him.

Daniel J. O'Keefe
25 Sunnyside Lane
Walnut Creek, California 94596
(415) 965-5418

Summary: Management Systems Analyst with 10-year successful track record in systems analysis, design, and programming for Fortune 500 client companies. Skillful problem solver with a strong foundation in computer sciences, electrical engineering, and mathematics.

Experience and Accomplishments

1985-Present   *Manager, Systems Design*
                The Berkeley Group, Oakland, California

Managed team of 20 programmers, systems designers, and computer engineers for management consulting firm with revenue of more than $10 million per year.
- Designed new systems, supervised installation and initial operation, including orientation and training of client personnel.
- Served as liaison with senior management of client firms to develop objective, review constraints, and recommend appropriate systems design.

1980-1985   *Systems Analyst*
                Electronic Solutions Corporation, Irvine, California

Top programmer and systems design expert with this innovative software development firm.
- Led the team developing Write-Right, a new word-processing program with grammar and spelling check features that has met with great market success.
- Redesigned ESC's order entry and shipping program, which resulted in 50% reduction in order-to-ship time.

Education

M.S., Computer Sciences, University of California at Irvine
B.S.E.E., California Polytechnic Institute, San Luis Obispo, CA, minor in mathematics

Honors

Graduated *summa cum laude*, Member *Phi Beta Kappa*

Publications

Master's Thesis: "Computer Programming for IBM Mainframes: A Systems Approach" published by University Press

Special Skills and Knowledge

IDMS/IDD, MVS INTERNALS, VM PERFORMANCE TUNING
Fluent in Spanish.

References

Provided on request, once mutual interest has been established.

PROBLEM RESUME: TEACHER/SPECIAL EDUCATOR

Hannah G. Allison
532 Blue Mountain Road, #7
Los Angeles, CA 90027
(2) (213) 665-5443 (H)
(714) 956-0811 (W)

(1) SPECIAL EDUCATOR

GENERAL BACKGROUND AND SKILLS

Eight years of experience in the field of special education: diagnosis and remediation of specific learning disabilities, development and implementation of (3) specialized curriculum for the mentally retarded; program modification for students with learning disabilities; kindergarten readiness testing. Strong organizational and communication skills.

EXPERIENCE

1981 To Present   Young Adult Center, Cerritos, CA
                      Program Resource Specialist
Responsibilities include development of curriculum, co-leading specialty groups,
(4) evaluating appropriateness of programming through the Utilization Review process, program compliance with state regulations. Developed systems for record keeping. Established student internship program with local colleges. Developed and carried
(5) out Program Evaluation system and Community Education/Outreach program. Responsible for writing and submission of grant proposals. Direct care with profound
(6) MR. multiple-handicapped adults.

1980-1981   Vista del Mar School, Los Angeles, CA
                      Special Teaching Assignment
Teacher for homebound profoundly retarded child. Responsibilities included develop-
(7) ing and carrying out daily program instruction and home management. Skills stressed included sustained eye contact, grasp, balance, awareness of sounds, touch, and self.

1980-1981   Johnston Central School, Los Angeles, CA
                      Teacher for Resource Room
(8) Responsibilities included development of program goals. Diagnosed and remediated specific disabilities; recommended program modifications for students in mainstream classes. Prepared students for RCT in reading, writing, and mathematics.
                                (9)
1976-1980   Los Angeles City Schools, Los Angeles, CA
                      Temporary and Substitute Teacher
(10) Long-term assignments in primary grades; developed teaching strategies and modified curriculum for disadvantaged children. Substitute in all grades in elemen- tary school and special education classes for the trainable and educable mentally retarded.

(11) CERTIFICATION   Special Classes of the Mentally Retarded and N-6
                        Permanent, California

EDUCATION   (12) M.S. Education; California State University, Dominguez Hills
                         B.S. Education & Behavioral Science, Pomona College

RESUME KEY FOR HANNAH ALLISON: TEACHER/SPECIAL EDUCATION

General Issues

Wordy and ambiguous. Unimpressive lead paragraph. Type format is weak. Visually, the resume needs more specific points of focus. Several key elements are missing.

1. Incorrect positioning of profession.
2. Omit work telephone number.
3. Lead paragraph does not point out accomplishments.
4. Needs action verbs.
5. Inconsistency in verb tenses.
6. Omit abbreviation. It was mentioned previously.
7. Same time frame. Combine format. Needs action verbs.
8. Same time frame. Combine format. Needs action verbs.
9. Assumes that reader understands abbreviation. Needs to be spelled out.
10. Omit position. Goes back too many years.
11. Combine with education.
12. Needs state.

HANNAH. G. ALLISON
532 Blue Mountain Road, #7
Los Angeles, CA 90027
(213) 665-5443

SUMMARY

Eight years of experience in special education. Strong diagnosis and remediation skills for specific learning disabilities. Developed and implemented specialized curriculum for mentally retarded. Created program modifications for students with learning disabilities.

EXPERIENCE

1981- Present   YOUNG ADULT CENTER, Cerritos, CA
Program Resource Specialist
Ensure compliance with state regulations for day treatment programs for 270 mild to profoundly mentally retarded handicapped adults. Design curriculum, co-lead specialty groups, and evaluate program validity from Utilization Review Process. Write and submit grant proposals. Establish student internship programs with local colleges. Create and implement Program Evaluation System and Community Education/Outreach Program. Administer direct care. Devise systems for record keeping.

1980-1981   JOHNSTON CENTRAL SCHOOL, Los Angeles, CA
Resource Room Teacher
Developed program goals and recommended program modifications for students in mainstream classes. Diagnosed and remediated specific disabilities. Prepared students for Regional Competency Test (RCT) in reading, writing, and mathematics.

VISTA del MAR SCHOOL, Los Angeles, CA
Special Teaching Assignment
Taught homebound profoundly mentally retarded child, after regular day's work. Established daily program of instruction and home management. Stressed skills in sustained eye contact, grasp, balance, awareness of sounds, touch, and self.

EDUCATION

Certification: Special Classes of the Mentally Retarded and N-6, Permanent, California

M.S. Education, California State University, Dominguez Hills

B.S. Education and Behavioral Science, Pomona College, Pomona, CA

(1)

RESUME
Heidi Chisholm

(2) Career Goal: Entry-level position in TV Production

Home Address: 634 Clevinger Street
Baltimore, MD 06346
(507) 231-3882

(3) Local Address: 10 S. Moore Street
Boston, Mass 02521
(617) 430-6731

(4) Born: March 3, 1972
Garden City, NY

Education: 1990-1992 Elementary Education major at the
(5) School of Education, North Central State
University
1992-1994 Senior at the School of Public
Communications: North Central State University
Graduate: B.S. Broadcasting and Film, 1994

High School: 1990 Graduate, East Baltimore
H.S.

1994:   Work at T.V. Graphics, distributing
equipment, at North Central State
University's School of Communications.

1994:   Audio Crew for North Central State University
Alumni Film.

1992:   School T.V. production presented at public
showing.

1992:   Volunteer work for North Central State's
closed-circuit radio station, WSBC, research
and on air.

Special Skills:
(6) Experienced in video equipment, including porta-paks,
studio cameras, mixing and switching boards, and slide
machine. Experience with audio console, including cart
machine, revox, and tape machines. Experienced in
cameras, viewers, and splicers.

References:
Will be furnished upon request.

RESUME KEY FOR HEIDI CHISHOLM: TV PRODUCER-ASSISTANT/ NEW GRADUATE

General Issues

Chisholm has no paid experience, but is long on credits through her university training. She needs to capitalize on them by listing her credits and explaining her assignments on the major productions. A modified targeted resume will work best.

1. Eliminate "Resume."
2. Change to "Job Target."
3. Omit "Local Address."
4. Omit date of birth.
5. Modify entire education section.
6. Edit to tighten.

HEIDI CHISHOLM
634 Clevinger Street
Baltimore, MD 06346
(507) 231-3882

JOB TARGET: Entry-level position in TV production.

EDUCATION: 1994: BA, Broadcasting and Film, North Central State University, Boston, MA.

CAPABILITIES:
* Television - Operate the following equipment: studio cameras, porta-paks, and switching panel.
* Television - Production Graphics.
* Radio - Operate the following equipment: audio console, tape machines. Handle cueing and mixing.
* Radio - Research and on air talent.
* Film - Operate the following equipment: Super 8 cameras, viewers, splicers, various projectors.

ACHIEVEMENTS: *TELEVISION
Produced and directed the following video productions:
"The Art of Batiking," "The Impossible Dream," "The Creative Process," and "Wildlife Conservation." Organized all aspects: scriptwriting, audio selection and placement, set design - including furniture building and prop acquisition, lighting design and crew, casting, making slides and cue cards, planning camera shots, angles, and composition.

*RADIO
Produced a tape demonstrating special effects including echo, reverberation, and speed distortion. Developed a 20-minute documentary: handled interviewing, narration, editing and splicing, and final taping.

*FILM
Produced and directed the following: "Everybody Is a Star," and "Love Is a Beautiful Thing." Handled camera work, editing, splicing, lighting, and soundtrack. Designed and produced all graphics.

REFERENCES: Furnished on request.

PROBLEM RESUME: URBAN RESOURCES MANAGER/FORESTRY

STEPHEN H. CAMPBELL
143 Thurston Lane
Oakland, CA 94591
(510) 642-9341

WORK EXPERIENCE:

UNITED STATES FOREST SERVICE ①   ② September 1990 - November 1991

Research Assistant
Led project in determining existing and potential tree growth throughout various California cities. The study compared and contrasted the amount of foliage growth of each city with the amount of other cities in different climatic regions. Included extensive use of maps and aerial photographs.

Assisted a landscape architecture firm in an urban forestry project for the city of Sacramento. Analysis of public tree growth and open space.

Helped organize and complete an in-depth study for the Department of Public Works in association with Trust for Public Lands, in frost damaged street trees for the city of San Francisco.

Worked for N.A.S.A., at Ames Research Center in an existing vegetation growth analysis for Northern California urban areas. Developed and implemented research procedure, mapped and digitized all study locations.

UNIVERSITY OF CALIFORNIA AT BERKELEY   ③ December 1989 - May 1990

Research Assistant
Set up and maintained laboratory experiments for Dr. Samuel Collier, a Landscape Architecture and Forestry Professor. Consisted of research and data collecting in various Urban Forestry projects. For example, took part in a three-week seedling experiment, analyzing the impact of various climate and soil conditions on seedling growth. Also assisted in Bay Area Urban Creek Study.

EDUCATION:

UNIVERSITY OF CALIFORNIA AT BERKELEY
Bachelor of Science, General Resources, May 1991
④

⑤ Classes covered subjects such as Plant and Soil Biology, Environmental Chemistry, Landscape Architecture, Forestry Ecology, Geology, Geography, Plant Pathology, Entomology, and Forestry Economics. Degree required participation in Summer Forestry Camp, an 8-week program in the Sierra Mountains that provided "field" experience for issues addressed in classroom.

⑥ Member, Phi Kappa Sigma Fraternity.

SPECIAL SKILLS:

IBM - LOTUS 1-2-3, WordPerfect
Stereoscope - a photographic 3-D imaging device
Digitizer - computerized boundaries of desired study areas
Arc/Infor, Arcedit - software program used with digitizer
Field Instruments - compass, clinometer, tree bore, altimeter, PH testing equipment, among others

INTERESTS:

(7) Reading, creative writing, football, baseball, golf, and backpacking.

RESUME KEY FOR STEPHEN H. CAMPBELL: URBAN RESOURCES MANAGER/FORESTRY

General Issues

Though he has excellent qualifications and experience for a recent graduate, his resume is too long and wordy. Rewriting it into a functional format focuses on his strengths, without emphasizing the relatively short amount of time spent on work projects.

1. No location of employment noted.
2. Omit months.
3. See #2
4. Explain nature of major.
5. Class format needs to be condensed.
6. Omit fraternity; no benefit to inclusion.
7. Omit "Interests"; unnecessary and adds to excess length.

STEPHEN H. CAMPBELL
143 Thurston Lane
Oakland, CA 94591
(510) 642-9341

EDUCATION:

BS General Resources Management (Forestry Degree: Emphasis, Urban Management)
UNIVERSITY OF CALIFORNIA AT BERKELEY   1991

Related Coursework:

Plant & Soil Biology	Environmental Chemistry
Landscape Architecture	Forestry Ecology
Geology	Geography
Plant Pathology	Entomology
Forestry Economics	Summer Forestry Camp for fieldwork

WORK EXPERIENCE:

VEGETATION RESEARCH: Led project to determine existing and potential tree growth throughout California cities, comparing and contrasting to cities in other climatic zones. Included extensive use of maps and aerial photographs. Developed and implemented research in existing vegetation growth for Northern California urban areas on behalf of NASA at Ames Research Center. Developed and implemented research procedures, mapped and digitized all study locations.

URBAN FORESTRY: Analyzed public tree growth and open spaces for City of Sacramento, CA. Helped organize and complete in-depth study on frost-damaged trees for City of San Francisco's Department of Public Works through Trust for Public Lands. Worked with Samuel Collier, UC professor of Landscape Architecture & Urban Forestry on climate and soil studies.

EMPLOYMENT:

1990 - 1991 Research Assistant, United States Forest Service, Oakland, CA
1989- 1990 Research Assistant, University of California at Berkeley

SPECIAL SKILLS:

* IBM - Lotus 1-2-3, WordPerfect
* Stereoscope - photographic 3-D imaging device
* Digitizer - computerized boundaries of desired study areas
* Arc/infor, Arcedit - software program used with digitizer
* Field Instruments - compass, clinometer, tree bore, altimeter, pH testing equipment, among others

REFERENCES SUPPLIED UPON REQUEST

PROBLEM RESUME: WORD PROCESSOR

Yvonne Daly
2356 Cassandra Blvd., #305
Detroit, MI 48202
(313) 823-6711

CAREER OBJECTIVE:

(1) A challenging position as a word processor, where I may utilize my skills and contribute while advancing with the company.

EDUCATION:

2/90 TO **MICHIGAN COMPUTER INSTITUTE**
PRESENT 20775 Greenfield
(2) Southfield, MI 48075 (313) 443-5400
CURRICULUM: Computerized Word Processing
SOFTWARE: DOS, DisplayWrite 4, WordPerfect 5.0, DTP
HARDWARE: IBM PC, manual and electronic typewriter.

8/85 TO 6/89 **NORTHERN HIGH SCHOOL**
9026 Woodward
Detroit, MI 48202 (313) 494-2626
(3) CURRICULUM: Business Skills; Typing 1-4
Computer Literacy, Basic Computer
(4) Received Diploma

EMPLOYMENT:

6/88 to 8/88 **TURNER CONSTRUCTION COMPANY**
(5) (summer only) 150 W. Jefferson, Suite 500 (7)
Detroit, MI 48226 (313) 596-0500
10/87 to 2/88 Began employment as a Accounting Clerk.
(6) (co-op only) Responsible for filing, light typing, errands, and
using the word processor.
(8) SUPERVISOR: David Allen

2/88 TO 6/88 **DETROIT BOARD OF EDUCATION**
(9) (co-op only) 9026 Woodward (10)
Detroit, MI 48202 (313) 494-2625
Began employment as a Office Aide.
Responsible for greeting guests, answering the
telephone, and sorting mail.
(11) SUPERVISOR: Mrs. Jessie Gibson-Eldrige

11/90 to 2/92 CROWLEY'S DEPARTMENT STORE
3031 W. Grand Blvd.
Detroit, MI 48202 (313) 874-5100
Cashier, customer service, restocking merchandise,
general cleaning (12)

(13) SUPERVISOR: Mrs. Sandy Schowdowsky

REFERENCES: Available upon request

271

General Issues

You'd expect a trained word processor to have a neat and tidy resume. Yvonne Daly does. It just doesn't have much else going for it. Daly is relying on her skills to land her that all-important entry-level job in her chosen vocation. Yet she omits all-important facts like her keyboarding speed, and includes unnecessary information like her past employers' addresses, phone numbers, and supervisors. With a functional format and a shift of emphasis to her skills and capabilities, she can move ahead with ease.

1. Omit objective.
2. Address and phone number are unnecessary.
3. See #2.
4. Incorrect terminology.
5. "Summer Only" irrelevant.
6. "Co-op Only" irrelevant.
7. Incorrect article.
8. Supervisor irrelevant.
9. See #6.
10. See #7.
11. See #8.
12. Omit "General Cleaning."
13. See #8.

YVONNE DALY
2356 Cassandra Blvd., #305
Detroit, MI 48202
(313) 823-6711

EDUCATION:

1990 To Present	Graduating 1992 with Certificate of Completion/Word Processing
	Michigan Computer Institute, Southfield, MI
1985 - 1989	Graduate, Northern High School, Detroit, MI
	Curriculum: Business Skills, Typing 1-4, Basic Computer

BUSINESS SKILLS:

* Keyboarding proficiency test passed @ 65 wpm

* Software: WordPerfect 5.0, DisplayWrite 4, DOS

* Hardware: IBM PC, manual and electronic typewriter

* Filing, mail sorting, personal and telephone reception

* Cashier, customer service, merchandise restocker

CAPABILITIES:

* Respond positively to challenge of deadlines

* Work cooperatively with a wide range of personalities

* Listen attentively to instructions so they don't need to be repeated

EMPLOYMENT:

1990 -1992	CROWLEY'S DEPARTMENT STORE, Detroit, MI
	Cashier and Customer Service Representative
1987-1988	TURNER CONSTRUCTION COMPANY, Detroit, MI
	Accounting Clerk
1988	BOARD OF EDUCATION, Detroit, MI
	Office Aide

REFERENCES: Available upon request.

Index

FREE RESUME CRITIQUE

Tear out this form and send it with your resume and self-addressed, stamped envelope to:

Resume Makeover Response
P.O. Box 34444
Los Angeles, CA 90034

Resume Status for: (write your name) _____

	Excellent	Good	Fair	Poor
I. First Impression				
1. Does it highlight your most significant achievements?	____	____	____	____
2. Does it read well?	____	____	____	____
3. Does experience fit objectives?	____	____	____	____
II. Content				
1. Logical sequence of events	____	____	____	____
2. Clear job descriptions	____	____	____	____
3. Quality of accomplishments	____	____	____	____
4. Quantity of accomplishments	____	____	____	____
5. Sentence construction	____	____	____	____
6. Sentence length	____	____	____	____
7. Understandable vocabulary	____	____	____	____
8. Vocabulary appropriate to target position	____	____	____	____
9. Format that fits goal	____	____	____	____
10. Balance of detail	____	____	____	____
III. Appearance				
1. Length	____	____	____	____
2. Typeface	____	____	____	____
3. Type size	____	____	____	____
4. Spacing	____	____	____	____
5. Paper quality and color	____	____	____	____

Resume Status for: _____

Explanation of numbers, margin notes, and marks on your resume:

1. Wordy—Reduce this section.
2. Expand.
3. Give more specifics.
4. Use more action words.
5. Change paragraph order.